A DICTIONARY OF
Scientific Units

A DICTIONARY OF
Scientific Units
Including dimensionless numbers and scales

H. G. JERRARD
BSc, PhD, FInstP

Reader in Physics
University of Southampton, UK
Formerly Professor of Physics
Oklahoma State University
Oklahoma, USA

and

D. B. McNEILL
TD, MSc, PhD, FInstP

Formerly Senior Lecturer in Physics
University of Southampton, UK

FIFTH EDITION

LONDON   NEW YORK
CHAPMAN AND HALL

First published 1963
by Chapman and Hall Ltd
11 New Fetter Lane, London EC4P 4EE
Second edition 1964
Third edition 1972
Fourth edition 1980
Fifth edition 1986
Published in the USA by
Chapman and Hall
29 West 35th Street, New York, NY 10001
© 1963, 1964, 1972, 1980, 1986
H. G. Jerrard and D. B. McNeill

Printed in Great Britain by
J. W. Arrowsmith Ltd, Bristol

ISBN 0 412 28090 6 (cased)
ISBN 0 412 28100 7 (Science Paperback)

British Library Cataloguing in Publication Data

Jerrard, H. G.
  A dictionary of scientific units:
  including dimensionless numbers and scales.
  —5th ed—(Science paperback; no.   210)
  1. Units—Dictionaries
  I. Title   II. McNeill, D. B.
  503'.21   QC82
  ISBN 0-412-28090-6
  ISBN 0-412-28100-7 Pbk

Library of Congress Cataloging in Publication Data

Jerrard, H. G.
  A dictionary of scientific units.

  (Science paperbacks; 210)
  Bibliography: p.
  Includes index.
    1. Units—Dictionaries. I. McNeill, D. B. (Donald
Burgess), 1911–   . II. Title. III. Series.
QC82.J4   1986   530.8'03'21   85-31431
ISBN 0-412-28090-6
ISBN 0-412-28100-7 (pbk.)

Preface to
the first edition

The intense specialization that occurs in science today has meant that scientists working in one field are often not familiar with the nomenclature used by their colleagues in other fields. This is particularly so in physics. This dictionary is designed to help overcome this difficulty by giving information about the units, dimensionless numbers and scales which have been used, or are still being used, throughout the world. Some four hundred entries are provided and these are supplemented by about five hundred references. The definition of each entry is given together with relevant historical facts. Where appropriate, some indication of the magnitude of each unit is included. Any scientific unit, which to the authors' knowledge has appeared in print, even if not universally adopted, has been listed. While it is too much to hope that there are no omissions, it is believed that there cannot be many and that this dictionary provides the most complete information of its kind available. The units are listed alphabetically and the references are numbered in sequence for each letter. In appendices are given a table of fundamental physical constants, details of standardization committees and conferences, a table of British and American weights and measures and conversion tables. The symbols and abbreviations used throughout the text are those recommended by the Institute of Physics and the Physical Society and by the British Standards Institution.

It is a pleasure to acknowledge the help given by numerous friends and colleagues and in particular that given by Dr J. R. Clarkson of the Royal South Hants Hospital; Mr C. H. Helmer of the Mechanical Engineering Department, Southampton University, Dr Peter Lane of the Iraq Petroleum Company; Miss D. M. Marshallsay of the Department of Economics, Southampton University; Mr R. E. Peroli of the Belfast Public Textile Testing House; Dr L. G. A. Sims, Professor of Electrical Engineering, Southampton University; Mr R. W. Watridge, the Southampton Borough analyst; Major H. W. H. West of the British Ceramic Research Association and by the Librarians and Staffs of the University and the Public Libraries at Southampton. Finally, we wish to thank Mrs H. G. Jerrard and Miss A. J. Tutte for typing the manuscript.

Department of Physics   H. G. JERRARD
University of Southampton   D. B. McNEILL
1963

Preface to
the fifth edition

Since the publication of the fourth edition in 1980 advances in technology have led to more precise values of the fundamental physical constants and a movement towards definitions of the fundamental units of mass, length and time based on atomic parameters. More precise definitions of some other units such as the candela have been approved by the international committees. These changes, together with the definitions of several new units have been included in this edition, the text of which has been revised and which now contains over 850 units and dimensionless numbers.

The authors wish to thank all those who have helped in this latest compilation by suggestion and kindly criticism and Margaret Wainwright who has had the difficult and tedious task of typing, retyping and copying the fragmented parts that arise from a text revision. At the time of going to press we believe this book to provide the most complete and up-to-date information of its kind available.

H. G. JERRARD   *1985*
Department of Physics
University of Southampton

The Mayor's Parlour
Fareham, Hants

D. B. McNEILL
Newtownards, Northern Ireland

Contents

Introduction

All non-electrical physical quantities may be defined in terms of mass, length and time which are the three fundamental mechanical units. Electrical and magnetic quantities generally require four units of which three are mass, length and time and the fourth can be some electrical or magnetic quantity such as current, permeability or permittivity. It is sometimes convenient to introduce temperature as an independent unit ranking equally with length, time and mass but if it is recognized that heat is of the same nature as energy, temperature may be defined in terms of the three fundamental mechanical units. It would appear, then, that all physical quantities can be expressed in terms of four units. The measurement of mass, length and time dates back to the dawn of history, whereas electrical and magnetic phenomena were not considered quantitatively until the middle of the nineteenth century.

In principle, measurement is finding an expression for a quantity in terms of a ratio of the quantity concerned to a defined amount of the same quantity. The defined amount can be chosen in an arbitrary manner – e.g. the yard – or it can have its origin in some natural phenomenon, such as the metre, which can be defined in terms of the wavelength of a selected line of the krypton emission spectrum. The former type of units are called arbitrary, the latter natural. At present the fundamental units of length and time are natural units whereas that of mass is arbitrary. Natural units are, in theory at least, capable of being reproduced anywhere at any time, whereas arbitrary units require the presence of a prototype. Once an arbitrary unit is defined it does not alter and neither would a natural unit once its real value is known, but in practice natural units have to be changed in size every time a discrepancy is found in the natural measurement from which the unit is derived. An example of this is given by the kilogram which was originally intended to be a natural unit representing the mass of a cubic decimetre of water. More accurate measurement at a later date showed the volume of water used was 1·00028 cubic decimetre, so the kilogram is a little too heavy. This could have

1

been remedied by altering the mass of the kilogram, but it was decided to keep the mass unchanged and to make the kilogram an arbitrary unit which is called the International Prototype Kilogram.

In general, units should be of such a size that the quantity being measured can be expressed in convenient figures. Thus the expenditure of a modern state may be given in units of millions of pounds but the cash in a child's money box is better counted in pence. It sometimes happens that the choice of a unit is not suitable to cover all applications in which case the original unit is multiplied by a suitable power of ten to give a derived unit. For example the unit of length is the metre but it is often better to give the distance between two towns in kilometres (10^3 m) and the size of bacteria in micrometers (10^{-6} m). A measurement should always be expressed as a number followed by the name of the unit concerned so that the magnitude and description of the physical quantity concerned are conveyed to all who require to know them. A number without a unit is meaningless unless it be a recognized dimensionless number or a ratio. A unit should always be called by its recognized name.

Systems of units

Any system of units defined in terms of the fundamental units of mass, length and time form an absolute system of units. The principle of absolute units was first proposed by K. F. Gauss (1777–1855) in 1832[1]. In 1851[2], he and Weber drew up a set of units based on the millimetre, the milligram and the second. This is known as the Gaussian system. Twenty-two years later[3] the British Association adopted the metric system but used the centimetre, the gram and the second as the fundamental units; these are often called the CGS units. The metric system recommended today is the International, or SI, system. This is based on the metre, kilogram and second. Engineers throughout the English speaking world generally use the foot, pound, second units. The history of the fundamental units is given briefly by Sir Richard Glazebrook in the 1931 Guthrie Lecture[4] and in some detail by R. W. Smith in the National Bureau of Standards Circular 593 entitled the Federal Basis for Weights and Measures[5]. The definitions of the seven SI base units are given in the National Physical Laboratory publication entitled *The International System of Units* (1977).

Electrical and magnetic quantities generally require a fourth term for their complete definition. In the CGS electromagnetic system, the units of which are also called e.m.u. or abunits, the fourth term is permeability and its value is taken as unity if the medium be a vacuum. In the CGS electrostatic system, in which the units are called e.s.u. or statunits, the fourth term is permittivity and this is considered to be unity in a vacuum. In the MKS system the fourth term is the permeability of free space and this has a value of $4\pi \times 10^{-7}$ henry metre^{-1} or 10^{-7} henry metre^{-1} according to whether a rationalized or unrationalized system is used. In SI units it is the ampere.

Atomic system of units

This system of units was suggested by Hartree in 1927[6] to reduce the numerical work in problems involving the atom. The unit of charge e is the

3

charge on the electron (160×10^{-21} coulomb), the unit of mass m is the rest mass of an electron (900×10^{-33} kg) and the unit of length a is the radius of the first Bohr orbit in the hydrogen atom (53×10^{-12} m). The unit of time is the reciprocal of the angular frequency $1/4\pi Rc$ (24×10^{-18} second), where R is Rydberg's constant and c the speed of light. The unit of action is $h/2\pi$, where h is Planck's constant. The unit of energy is $e^2/a^2 = 2Rhc$, which is the potential energy of unit charge situated at unit distance from a similar charge and which is also equal to twice the ionization energy of the hydrogen atom. Shull and Hall[7] suggest that since e^2/a^2 can be written as $4\pi^2 me^4/h^2$, this quantity should be used as a unit of energy which they call the hartree. Other somewhat similar systems of atomic units have been developed in the past 60 years one of the most recent being by McWeeny[8] who used the electronic charge, the Bohr radius, the unit of action and the permeability of free space as his base units.

British Imperial Units

The majority of countries in the British Commonwealth have used imperial units which are based on the pound, the yard and the second. The standards for the first two are kept in London, the imperial standard pound being equal to about 0·453 592 43 kg and the imperial standard yard to approximately 0·914 399 m. The Weights and Measures Act of 1963 removed the independent status of these two quantities which are now based on the kilogram and the metre, the pound being defined as exactly 0·453 592 37 kg and the yard to 0·9144 m precisely. The unit of time, the second, has always been the same in both systems and was redefined by the 13th CGPM (1968) as the duration of 9 192 631 770 periods of radiation corresponding to the transitions between two hyperfine levels in the ground state of the caesium 133 atom.

CGS, e.m.u., e.s.u., system of electrical units

The force F between two magnetic poles of strength m_1 and m_2 placed a distance d apart in a medium of permeability μ is given by $F = m_1 m_2/\mu d^2$. If F, μ and d be each unity and if $m_1 = m_2 = m$, then m_1 and m_2 are poles of unit strength. Units based on this definition of m are known as electromagnetic units (e.m.u.). The quantities defined by these units are generally of an inconvenient size for practical work, so units known as practical units are used. The latter may be obtained by multiplying the e.m. unit by a suitable conversion factor. In 1903[9] the prefix ab was suggested to denote the unit concerned is a CGS electromagnetic unit, thus 1 abvolt = 10^{-8} practical volts, 1 abampere = 10 practical amperes. The suggestion met with little response at first but in recent years some authors have started using this notation[10].

The force F between two charges q_1 and q_2, a distance d apart in a medium of permittivity ε, is given by $F = q_1 q_2/\varepsilon d^2$. If F, ε and d be unity and if $q_1 = q_2 = q$, then q_1 and q_2 are unit charges. Electrostatic units (e.s.u.) are based on this value of q. Like the e.m. units electrostatic units are not of convenient size for practical work and are generally replaced by practical units for everyday electrical measurements. In recent years the prefix stat has sometimes been used to denote electrostatic units, thus 1 stat volt $= 300$ practical volts, 1 stat ampere $= (1/3) \times 10^{-9}$ ampere. This prefix is an abbreviation for abstat which was proposed for electrostatic units at the same time as ab was suggested for electromagnetic units.

Ab units and stat units are connected by the relationship $\mu\varepsilon = 1/c^2$, where c is the velocity of light in centimetres second^{-1}. Thus the ratio (ab unit/stat unit) for the primary units is equal to the velocity of light, or its reciprocal, viz. abampere/statampere $= c$ and abvolt/statvolt $= 1/c$. The ratio ab/stat for secondary units is obtained by considering each of the primary units concerned thus

$$\frac{\text{ab farad}}{\text{stat farad}} = \frac{\text{ab coulomb}}{\text{ab volt}} = \frac{\text{stat volt}}{\text{stat coulomb}} = c^2$$

The inconvenience of having three systems of electrical units, ab units, stat units and practical units has been overcome by the introduction of the metre, kilogram, second, ampere units (MKS). In this system, the practical units have the same value as the theoretical ones, which themselves require no modification for use in either electromagnetic or electrostatic problems.

Gravitational system of units
In these units the three fundamental quantities are the unit of weight, the unit of length and the unit of time[11]. In the British system these are the pound weight, the foot and the second. In the metric system the units are either the gram weight, the centimetre and the second (CGS) or the kilogram weight, the metre and the second (MKS or SI). In both systems the unit of weight is the fundamental unit of mass when weighed in a standard gravitational field for which the value of g, the acceleration due to gravity, is known. The International Commission of Weights and Measures agreed in 1901 that for gravitational units g should have a value of $32 \cdot 1740$ ft sec^{-2} or $980 \cdot 665$ cm sec^{-2}. Some 70 years later the 61st CIPM (1972) recommended that, whenever precision measurements involving gravitational forces have to be made, the relevant local value of g used should be that obtained from the International Gravity Standardization Network (IGNS – 71).

Heaviside–Lorentz system of units
These are CGS units in which the force between two magnetic poles m_1 and m_2, a distance d apart in a medium of permeability μ is given as $m_1 m_2/4\pi\mu d^2$. An analogous equation $q_1 q_2/4\pi\varepsilon d^2$ gives the force between two charges q_1

and q_2 in a medium of permittivity ε. The Heaviside–Lorentz units were the earliest rationalized units; they were proposed by Heaviside[12] in 1883 and used by him in a classical paper on electrical theory published nine years later[13].

International system of units
See **SI units**, p. 7.

Kalantaroff units
These were proposed by P. Kalantaroff in 1929[14]. The fundamental quantities on which they are based are the metre, second, weber and unit charge. Some advantages are claimed for the units in electrical and magnetic calculations but for most physical problems they are too cumbersome. Thus mass is expressed as:

$$\text{charge weber metre}^{-2}\text{ second}^{-1}.$$

Ludovici system of units
A system of units proposed in 1956[15] in which the fundamental quantities are the free space values of the gravitational constant, permeability, permittivity and charge. These values give a unit of length equal to $4 \cdot 88 \times 10^{-36}$ m and a unit of time equivalent to $16 \cdot 3 \times 10^{-45}$ second.

MKS system of electrical units
The CGS electromagnetic and electrostatic units are somewhat inconveneint when calculating certain electrical properties such as inductance and capacitance as certain factors involving powers of ten or the figures 3 and 9 have to be introduced to derive practical units from those obtained from theoretical considerations. As early as 1873 Clerk Maxwell[16] showed that practical electromagnetic units could be substituted directly in the fundamental theoretical equations if the unit of length were taken as the earth's quadrant (10^7 m), the unit of mass as 10^{-11} gram, the permeability of a vacuum μ_0 as unity and the unit of time as the second. Thirty years later G. Giorgi[17] pointed out that if the unit of length be taken as the metre, the unit of mass as the kilogram, μ_0 as $4\pi \times 10^{-7}$ and the unit of time remain unchanged as the second, then the practical units could be used directly in both the electromagnetic and the electrostatic systems. Furthermore, the introduction of 4π into the value of μ_0 meant that most of the electrical units would be rationalized, i.e. the factor 2π would occur if the system were cylindrically symmetrical and 4π if spherically symmetrical. A disadvantage of the Giorgi system is that the difference between magnetic induction (B) and field strength (H) can no longer be ignored as it is in the CGS system in cases where the permeability is unity, such as when the medium is air. Similarly the

difference between the electric field (E) and the electric displacement (D) cannot be neglected as it is when the permittivity is unity.

The Giorgi or MKS system attracted little attention until about 1935[18] but after this interest in them increased and in 1948[19] the 9th International Conference on Weights and Measures adopted the MKS definition as their definition of the ampere and recommended the fourth unit in the rationalized MKS system be $4\pi \times 10^{-7}$ henry metre^{-1}, a term which is known as the permeability of free space. The ampere is now defined as the steady current which, when maintained in two parallel conductors of infinite length and of negligible cross-section one metre apart in a vacuum, produces between the conductors a force equal to 2×10^{-7} MKS units of force per metre length.

OASM units
A system of units proposed in 1945[20] in which the ohm, ampere, second and metre are the fundamental quantities.

SI units
The Système International d'Unités (designated SI in all languages) is a *redefinition* of the MKS system. The seven base units are the ampere, candela, kelvin, kilogram, metre, mole and second[21]. These are of convenient size for normal practical work and may be used to give a rationalized coherent system of units in which the magnitude of any physical quantity may be expressed. The units have been evolved over the years from the rationalized MKS units; six were finally approved by the 11th CGPM (1960) and the seventh, the mole, by the 14th CGPM (1972). By the mid-1970s they had been adopted as the legal units for trade in over 30 countries and have been recommended for use by scientists throughout the world. The advantages of the system are numerous, thus the Joule – which is the unit of energy – may be expressed either as newton × metre or watt × second and, in electrical and magnetic work, there is now no necessity to use special electrostatic and electromagnetic units with the resulting confusion when it is necessary to change from one to the other. Older scientists will probably find it difficult to accommodate themselves to the new system for many of the old familiar values have gone, e.g. density of water is now 1000 kg m^{-3} and the acceleration due to gravity is 9·81 m sec^{-2}. They may find some solace, however, in the loophole left open to them by the National Bureau of Standards which, while recommending the use of the SI system, finishes its exhortation by stating 'except when the use of these units would obviously impair communication or reduce the usefulness of a report to the primary recipients'.

Stroud system of units
These were devised by Professor W. Stroud of Leeds about 1880[22] to give

engineers a set of absolute units based on the pound, the foot and the second and in which the distinction between mass and weight was emphasized. Stroud used capital letters for forces and small letters for masses, thus one Pound could accelerate one pound by 32 feet per second per second.

TMS units
These are a metric system based on the tonne, metre and second.

A

Abampere (abcoulomb, abfarad, abohm, abvolt)
The prefix ab- denotes the CGS electromagnetic system of units, e.g. abampere is the unit of current in the CGS system. (*See* **CGS units,** page 4.)

Abbe
A name suggested in 1973 for an SI unit of linear spatial frequency[1]; it has the dimensions of $Hz\ m^{-1}$ and is named after Ernst Abbe (1840–1901) an eminent authority on optical instruments.

Acoustic comfort index (ACI)
An arbitrary unit suggested in 1951[2] to indicate the noise in the passenger cabin of an aircraft. The figure $+100$ represented ideal conditions, -100 was intolerable and zero indicated the noise was just tolerable.

Acoustic ohm
The acoustic ohm is the unit of acoustic impedance[3]. The acoustic impedance of a surface is defined as the ratio of the effective sound pressure averaged over the surface (i.e. pressure/area) to the effective volume velocity through it. This ratio is a complex number. Volume velocity is defined as the rate of flow of the medium perpendicular to the surface. A surface has an acoustic impedance of one ohm when unit effective pressure produces unit velocity across it. The CGS acoustic ohm has dimensions of dyne second cm^{-5}, whereas the dimensions of the MKS unit, called the MKS acoustic ohm[4], are newton second $metre^{-5}$. The baffle of a loudspeaker has generally an acoustic impedance of the order of several hundred MKS acoustic ohms. The idea of applying Kirchhoff's electrical circuit procedures to solve acoustical problems was suggested by Webster[5] as early as 1919 but the acoustic ohm was first used by Stewart[6] in 1926.

Acre
A unit of area equal to 4840 square yards. It was first defined in England in

9

the reign of Edward I (1272–1307) and is reputed to be the area which a yoke of oxen could plough in a day.

At one time Ireland and Scotland used a different acre from that employed in England, thus one English (Imperial) acre $= 0.82$ Scottish acres, $= 0.62$ Irish acres, $= 0.78$ Cunningham or Plantation acres; the last mentioned was used mainly in North East Ireland whereas elsewhere in that country areas were measured in Irish acres[7]. One acre $= 0.40467$ hectares.

Alfven number (Al)

A dimensionless number characterizing steady fluid flow past an obstacle in a uniform magnetic field parallel to the direction of flow. It has a partial analogy to the Mach number. The Alfven number is given by $vl(\rho\mu)^{1/2} B^{-1/2}$ where v is the velocity of flow, l is length, ρ is density, μ is permeability and B is magnetic flux density. It is named after H. Alfven, the Swedish astrophysicist, who introduced the term magnetohydrodynamics.

Amagat units

These are units of volume and density used in the study of the behaviour of gases under pressure. The unit of volume is taken as being the volume occupied by a gram mole of the gas at unit pressure and 273·16 K and thus for an ideal gas the Amagat volume is 22·4 litres. The Amagat density unit expresses the density of gas in gram moles per litre, one density unit being equal to 0·0446 gram mole litre^{-1} at unit pressure. For both units, unit pressure is taken as one standard atmosphere. The units are named after E. H. Amagat (1841–1915) who studied the effect of high pressures on gases. Amagat units have been used extensively in Holland since the time of J. D. van der Waals (1837–1923) but were not used in England until 1939[8].

American run

A unit used in the textile industry for describing the length per unit mass of a yarn. (*See* **Yarn counts.**)

Ampere (A)

The 9th Meeting of the International Weights and Measures Congress[9] in 1948 defined the ampere as the intensity of the constant current which, when maintained in two parallel straight conductors of infinite length and of negligible cross-section placed one metre apart in a vacuum, produced between them a force equal to 2×10^{-7} MKS units of force per metre length. This unit, known as the absolute or SI ampere, replaced the international ampere which had been defined in 1908. The international unit, which was basically similar to the ampere introduced at the first meeting of IEC in 1881, was defined as the unvarying current which deposited 0·00111800 grams of silver by electrolysis from a silver nitrate solution in one second[10]. It was discovered sometime later that the international ampere was smaller than the absolute ampere[11] (1 international ampere = 0·99985 absolute amperes) and it was replaced by the absolute unit in 1948.

The necessity for a practical unit of current which should have a recognised name was suggested originally by Sir Charles Bright and Latimer Clark[12]. They proposed it should be some multiple of the fundamental metric unit which was based on the strength of the magnetic field produced by a current in a conductor of specified dimensions. The name suggested by them was the galvat, but this title was never used. Instead the practice started at Chatham barracks[13] of calling it the weber was more popular and the British Association designated the weber as the practical unit of current at their 1881 meeting[14]. The value assigned to it was 10^{-1} CGS units of current. Later in the same year the unit was renamed the ampere, after A. M. Ampere (1775–1836) one of the pioneers of electrodynamics, at the first meeting of the IEC in Paris. Previously to this some English writers[15] were already using the ampere instead of the weber as the practical unit of current.

Ampere turn (At)

The ampere turn is defined as the product Ni, where N is the total number of turns in a coil through which a current of i amperes is passing. It is often used as the unit of magneto-motive force. Ampere turns are also used in the description of the electrical circuits of electro-magnets. The term was first used in 1892[16].

Ångström (Å)

A unit of length equal to 10^{-10} m. It is used in atomic measurements and for the wavelength of electromagnetic radiation in the visible, near infra-red and near ultra-violet regions of the spectrum. The unit is derived from the red emission line of the cadmium spectrum which has an internationally agreed wavelength of 6438·4696 Å in dry air at standard atmospheric pressure at a temperature of 15°C and containing 0·03 per cent carbon dioxide by volume.

The unit was introduced by the International Union for Solar Research in 1907[17]. It was named after A. J. Ångström (1814–74), the Scandinavian scientist who used units of 10^{-10} m to describe wavelengths in his classical map of the solar spectrum made in 1868. The ångström was not confirmed as a unit of length by the International Congress of Weights and Measures until 1927. For over half a century the ångström was equal to $1·0000002 \times 10^{-10}$ m but when the metre was defined in terms of the wavelength of krypton in 1960[18] the ångström became equal to 10^{-10} m exactly.

The ångström is sometimes called a tenth metre.

Apostilb (asb)

An apostilb is the luminance (brightness) of a uniformly diffusing surface which emits one lumen per square metre. It is equivalent to the luminance produced by $1/\pi$ candela or 10^{-4} lambert. Its use was authorized in 1935[19] but it is not recommended for scientific work.

Apothecaries weights
See Appendix 3.

Arcmin
A term sometimes used instead of minutes of arc.

Are (a)
A metric unit of area equal to 100 square metres. It was introduced, along with the other metric units, at the time of the French Revolution and its first appearance in an English text was in 1810[20]. It was approved by the CIPM in 1879 and is used today in agriculture on the mainland of Europe.

Assay ton (AT)
In the assaying of gold and silver the result is expressed in troy weight ounces of metal per ton of ore. This result is the same numerically as the mass of noble metal expressed in grams per x kg of ore. The quantity x is known as the assay ton and has values of $29 \cdot 19 \times 10^{-3}$ kg and $32 \cdot 67 \times 10^{-3}$ kg for the short (2000 lb) and for the long (2240 lb) tons respectively. The assay ton has been in use for over a century.

Astronomical unit (AU)
A unit for describing planetary distance. It is approximately the mean distance between the sun and the earth. It was recognized at the first meeting of the International Astronomical Union in May 1922[21] and its present value is $149 \cdot 597870 \times 10^{9}$ m[22].

Atmo-metre
This is based on Dalton's law of partial pressures. As there are $2 \cdot 687 \times 10^{-25}$ molecules in unit volume of gas at s.t.p., the pressure exerted by these is the same as that exerted by a column of gas 1 metre high. This height is known as an atmo-metre, or metre-atmosphere; thus if the partial pressures of two gases be x and y atmo-metres, every cubic metre of the mixture will contain $2 \cdot 687\, x \times 10^{25}$ and $2 \cdot 687\, y \times 10^{25}$ molecules of each gas and their partial pressures will be in the ratio $x:y$.

Atomic mass unit (amu)
The masses of atoms and molecules are generally given in atomic mass units. These units are based on a scale in which the mass of the carbon isotope C_6^{12} is taken to be 12. This makes one atomic mass unit equivalent to

1.660531×10^{-27} kg. The unit is sometimes called a dalton. Atomic masses were originally given as atomic weights on a scale where the mass of the hydrogen atom was unity. In 1885 Ostwald suggested if atomic weights were expressed on a scale in which the mass of oxygen was 16, more of the elements would have integral numbers for their atomic weights. The discovery of oxygen isotopes led to the adoption of two atomic weight scales[23]. The one used by chemists was based on the figure 16 representing the average mass of the oxygen atom in its natural state ($O^{16}:O^{18}:O^{17}$ in the abundance $506:1:0.204$), whereas that used by the physicists considered the oxygen isotope O^{16} as the basic unit on their scale. The ratio of atomic mass on the physical scale to atomic mass on the oxygen chemical scale[24] was 1.000272 ± 0.00005.

In 1960[25] the International Union of Pure and Applied Physics followed the proposal made a year earlier by the International Union of Pure and Applied Chemistry that all atomic weights should be based on the C_6^{12} scale. This enabled more isotopes to have integral mass numbers than would have been possible on the oxygen scale. This reduced the values given on the previous chemical scale by a factor of 1.000043 and changed the values of the Faraday and Avogadro's constant by a similar amount. Values on the physical scale can be converted to the new scale by multiplying by 0.999685.

Atomic number

The atomic number was introduced by J. A. R. Newlands[26] (1837–98) in 1865 to describe the position of an element in the periodic table. The work of Rutherford and Moseley (1913) showed that the atomic number also indicated both the number of electrons in an atom and the number of positive charges in the nucleus[27]. The values of the atomic numbers of the elements so far discovered lie between one for hydrogen and 103 for lawrencium.

Avogadro constant or number (N or N_A)

The Avogadro constant is the number of atoms in a gram-atom or the number of molecules in a gram-molecule. It is named after Amadeo Avogadro (1776–1856) an Italian who in 1811 introduced the famous hypothesis. The value is given in Appendix 1.

Avoirdupois weights

See Appendix 3.

B

Bailling
A specific gravity unit. (*See* **Degree (hydrometry).**)

Ball
See **Zhubov scale.**

Balmer
A name suggested in 1951 for the unit of wave-number[1]. It represents the number of waves in a centimetre and has the dimensions of cm^{-1}. The unit is named after J. J. Balmer (1825–1898), an early spectroscopist.

Bar
A unit of pressure equal to 10^5 N m^{-2} (10^6 dyne cm^{-2}). The unit in its present form was suggested by Bjerknes[2] in 1911 and was first used in weather reporting in 1914[3]. A bar is equal to 750·062 torr.

Two pressure units called bar were in use during the first half of the present century. One gave pressure in dyne cm^{-2}, the other in 10^6 dyne cm^{-2}. The former was used in acoustics but was replaced by the 10^6 dyne cm^{-2} bar in 1951[4]. Acoustic pressure of the order of one dyne cm^{-2} is now described by the microbar.

There is no recognized abbreviation for the bar, but the abbreviations mb and μb are used for the millibar (10^{-3} bar) and the microbar respectively[5].

Barad
A former unit of pressure equal to one dyne per square centimetre. It was proposed and named by the British Association in 1888[6].

Barn (b)
The unit of nuclear cross-section equal to 10^{-28} m^2 per nucleus. The radius of a nucleus is of the order 10^{-14} m, which leads to a figure of the order of 10^{-28} m^2 for the cross-sectional area. The unit gives a measure of the

15

probability of a particular nuclear process (e.g. absorption, fission, scatter) occurring when nuclear projectiles pass through matter by giving the effective target area of the bombarded nucleus for that particular area. The cross-sectional areas vary from 10^4 barn for certain nuclear reactions to 10^{-11} barn for electron bombardment experiments. The name barn was made up by H. G. Holloway and C. P. Baker in Chicago in 1942[7]. They used it originally as a code word to describe the probability of certain reactions. It is possible that the authors had the idea the barn provided a target sufficiently large so that even a temporary war-time rifleman could not miss it. In July 1976, the Council of Ministers of the EEC proposed the abolition of the barn and this has now become effective by writing the area as 100 fm^2.

Barrel

The barrel has been in use in England as a measure of volume since 1433 when it was decreed that every city, borough and town in the realm would keep a balance, a set of weights and a common starred barrel for the use of merchants. Today the barrel is used as a liquid measure, its volume depending upon the substance being stored, e.g. for alcohol it is 189 litres and 159 litres for petroleum; the latter is also equal to 42 gallons (US) or 35 gallons (British). The barrel is also used as a grain measure, one barrel of flour weighing 196 pounds and that of beans 280 pounds.

Barye

The barye was established in 1900[2] as a pressure unit equal to 10^5 Nm^{-2}. Sometime later the term was used by some experimenters as the CGS unit of pressure, i.e. one dyne per square centimetre. The name barye is not recommended by either the American Standards Association[4] or the British Standards Institution[8]. The unit is now called the bar (q.v.).

Base box

A unit of thickness used in plating in which the thickness is expressed in terms of the mass in pounds deposited per base box. A base box is taken to be the area formed by 112 plates each of size 10 inch × 14 inch, i.e. 31 360 square inches.

Baud

The baud is a unit of telegraph signalling speed; one baud is equal to one pulse per second. The unit was proposed at an International Telegraph Conference in Berlin in 1927. It is named after J M. E. Baudot (1845–1903), the French telegraph engineer who made the first successful teleprinter.

Baumé

A specific gravity unit. (*See* **Degree (hydrometry).**)

Beaufort scale

A wind scale introduced by Admiral Sir Francis Beaufort (1774–1857) in which numbers represent the wind velocity[9]. The scale is first mentioned in the admiral's diary in January 1806 and its general use was suggested in an article in the *Nautical Magazine* in 1832. The Navy adopted the scale in 1838 and the Board of Trade have used it since 1862.

The scale is given in Table 1.

TABLE 1. *The Beaufort wind scale*

Number	Description	Wind speed (miles/hour)
0	calm	<1
1	light air	1–3
2	light breeze	4–7
3	gentle breeze	8–12
4	moderate breeze	13–18
5	fresh breeze	19–24
6	strong breeze	25–31
7	moderate gale	32–38
8	fresh gale	39–46
9	strong gale	47–54
10	whole gale	55–63
11	storm	64–72
12	hurricane	>72

The number on the Beaufort scale is sometimes called the Beaufort number B and is related to the wind velocity V in miles per hour by the empirical formula $V = 1.87\,B^{3/2}$.

The Beaufort scale is represented on weather maps by a wind arrow which indicates the direction of the wind. Each feather on the arrow indicates two points on the Beaufort scale so that, for example, an arrow with one and a half feathers signifies a force three wind[10].

Admiral Beaufort also devised a letter code to indicate the type of weather. His weather code was first published in 1833 and today it is used everywhere throughout the world for conveying weather information.

Becquerel (Bq)

The becquerel is the SI unit of radioactivity[11]; 1 becquerel represents one disintegration, or other nuclear transformation, per second, i.e. 1 curie = 37×10^9 becquerel. It is named after A. H. Becquerel (1852–1908) the French physicist who discovered radioactivity in 1896. The unit was approved by the 15th CGPM in May 1975.

Bel

A number used mainly in English-speaking countries to express the ratio of two powers as a logarithm to the base ten. It is defined as $N = \log_{10} P_1/P_2$, where N is the number in bels and P_1 and P_2 represent the values of the powers. The unit is named after the telephone pioneer Alexander Graham Bell (1847–1922). The unit was first used by the Bell telephone network in USA in 1923[12] when it replaced the standard cable as the unit describing the attenuation on a telephone line; it was originally called the TU or transmission unit (q.v.) and was renamed bel in 1923; the name was given international recognition in 1928[13]. The size of the bel makes it a somewhat inconvenient unit so the decibel is frequently used instead: this, as its name implies, is equal to 1/10 bel. In continental Europe the neper is used instead of the bel.

Benz

A name suggested for the SI unit of velocity (m s^{-1}) but it has not met with general acceptance. It is named after Karl Benz (1844–1929) of Mannheim, who built the first motor car and drove it in Munich in 1886.

Beranek scale

An arbitrary noise scale used in acoustics. (*See* **Subjective sound ratings**.)

Bes

One of the names suggested for the gram by Polvani[14] in 1951. Its merit seems to be that it can be confused with no other quantity except that of the name of the well meaning, but rather low powered, Egyptian deity.

BeV

An abbreviation often used in the USA to express energy in units of 10^9 electron volts but the abbreviation GeV is preferred. (*See* **Electron volt**.)

Bicron

A unit of length equal to 10^{-12} m; the name, like that of the micron, is not recommended. (*See* **Stigma**.)

Binocular numbers

The first of the two numbers generally stamped on a pair of binoculars indicates the magnification, the second gives the diameter of the object glass in millimetres. Thus 8 × 30 indicates a magnification of 8 and 30 mm as the diameter of the object glass. This system of describing binoculars was introduced between the two World Wars. The product of the diameter of the object glass and the magnification is called the twilight capacity.

Biot (Bi)

The name used by the SUN Committee for the unit of current in the electromagnetic CGS system of units in 1961. The biot is that constant current intensity which, when maintained in two parallel infinitely long rectilinear conductors of infinite length and of negligible circular section, placed at a mutual distance of 1 centimetre apart *in vacuo*, would produce between these conductors a force of 2×10^{-3} newton per metre length. One biot is equal to ten amperes. The unit is named after the French physicist J. B. Biot (1774–1862).

Biot

A unit used for measuring rotational strength in substances exhibiting circular dichroism. Circular dichroism is the ability of a material to absorb left- and right-handed circularly polarized light and is expressed by the difference $\Delta\varepsilon$ of the absorption coefficients which varies with wavelength λ. If a curve is drawn of $\Delta\varepsilon$ against λ the total circular dichroic intensity is given by the area under the curve. The rotational strength R (CGS units) is obtained from the relationship

$$R = \frac{3 \times 10^3 \, hc}{8\pi^3 \, N \log e} \int \frac{\Delta\varepsilon \mathrm{d}\lambda}{\lambda}$$

where c is the velocity of light, h is Planck's constant and N is Avogadro's number. Some authorities use 32 instead of 8. Values of R are often expressed in biots where one biot is 10^{-40} CGS. The unit is named after J. B. Biot who discovered optical rotatory dispersion; the unit was proposed by Liehr in 1963[15].

Biot or Nusselt number (Nu)

This is a non-dimensional coefficient used in fluid dynamics to describe the heat transfer between a moving fluid and a solid surface. The number Nu may be derived from the relationship

$$\mathrm{Nu} = \frac{l}{T_1 - T_2}\left(\frac{\partial T}{\partial x}\right)_{x \to 0} = \frac{hl}{k}$$

where l denotes length, T_1 and T_2 are the temperatures of the surfaces with and without heat transfer, $(\partial T/\partial x)$ represents the temperature gradient normal to the solid surface, h is the coefficient of heat transfer and k is the thermal conductivity[16].

The Nusselt number (Nu^*) used in mass transfer problems is given by

$$\frac{m}{tA} \frac{l}{\rho D}$$

where m is the mass transferred across an area A in time t, ρ is the density and D is the diffusion coefficient.

The number was named the Biot number after J. B. Biot (1774–1862), the first man to express the laws of convection in a mathematical form. In 1933 Grober and Erk[17] proposed the number be called the Nusselt number to commemorate the German engineer W. Nusselt who derived the number in 1905. The American Standards Association approved the name Nusselt in 1941[18].

Bit

The binary unit of information used in digital computers; it represents the smallest piece of addressable memory within a computer. The binary system of notation uses only two symbols, zero (0) and unity (1), to represent specific values. These two symbols are properly called the 0 bit and 1 bit but are commonly called bits. The name, which is derived from the first and last two letters of the phrase binary digit, came into use in the early nineteen fifties.

Blink

This is an arbitrary unit of time equal to 10^{-5} day; thus 1 blink = 0·8640 second or 1 second = 1·157 blink.

Blondel

Moon[19] suggested in 1942 that the term luminance of a source be replaced by the helios of a source and that the unit be called the blondel after A. E. Blondel (1863–1938) the French physicist who established the present system of coherent photometric units which were accepted at the Geneva IEC meeting in 1896. 1 blondel is equal to π times the lumens per square centimetre per steradian or to 1 apostilb (q.v.).

Board of Trade unit

An electrical unit representing the energy dissipated by one kilowatt in an hour. (*See* **Kilowatt hour**.)

Bohr radius

The radius of the first Bohr orbit of the hydrogen atom is approximately 53×10^{-12} m and it was proposed as a unit of length by Hartree in 1928[20].

Bole

The name suggested by the British Association in 1888[6] for the CGS unit of momentum (g cm s^{-1}). The name has never been used.

Boltzmann constant

See Appendix 1.

Bougie-decimale

A French photometric unit defined in 1889[21]. The bougie-decimale was equivalent to 0·96 international candles. It is now obsolete, having been displaced by the candela (q.v.).

Brewster (B)

The Brewster is a unit in photo-elastic work in which the stress optical coefficient of a material is expressed. The relative retardation R produced between the components of a linearly polarized light beam, when it passes through a medium made doubly refracting by a compressive or tensile stress P is given by $R = CPl$, in which l is the thickness of the medium and C is the stress optical coefficient. A brewster is the stress optical coefficient of a material in which a stress of 10^5 Nm^{-2} produces a relative retardation of one ångström when light passes through a thickness of one millimetre in a direction perpendicular to that of the stress. It is equivalent to 10^{-14} m^2 N^{-1}. When l is measured in m and P in kg m^{-2}, the unit of C is 1/98·1 brewster, whereas if l is in inches and P in lb in^{-2}, C is in units of 1/1·752 brewster.

Engineers often use the material fringe value F or the model fringe value M as a unit instead of the brewster. F, given as lb in^{-1} per fringe, is $\lambda/1·1752C$, where C is in brewsters and λ is the wavelength in ångströms. It is thus the stress required to produce a retardation of λ in an inch thick sample. $M = F/t$ and is expressed as lb in^{-2} per fringe in^{-1}. American workers often use shear fringe value which is $F/2$.

The stress optical coefficient of glass varies from $4·5B$ to $-2B$, the sign indicating whether it behaves as a positive or negative doubly refracting medium. The brewster was named after Sir David Brewster (1781–1868), who formulated the stress optical law. It was first used in 1910 by Filon[22].

Brieze

One of the names suggested in 1951 by Polvani[14] for the gram. It is derived from the Greek word for heavy.

Brig

A name for a unit to express the ratio of two quantities as a logarithm to the base ten, i.e. it is analogous to the bel, but the latter is restricted to power ratios. The brig was suggested in 1934[23] and is named after Henry Briggs (1561–1630), who prepared the first table of common logarithms in 1616[24].

Bril

A unit of subjective brightness[25]. The bril scale was drawn up by Hanes in 1949, but it is somewhat indefinite due to the different reactions of observers to brightness.

Brinell number
A hardness number (q.v.).

British thermal unit (Btu)
The British thermal unit is the energy required to raise the temperature of one pound of water through one degree Fahrenheit. As this depends upon the temperature of the water it is usual to include some indication of the temperature range to which the particular Btu refers, thus the Btu_{39} refers to the energy required to raise the temperature of the water from $39°$ to $40°F$. The value is 1059·52 joules. The Btu (mean) is defined as the 180th part of the quantity of heat required to raise one pound of water from $32°$ to $212°F$; it is equal to 1055·79 joules. The Btu_{IT} is that given in the *International steam tables* and is equal to 1055·06 joules. One Btu_{IT} per pound is equal to 2326 joules per kilogram.

The British thermal unit was the unit used by Joule in his original work on the relationship between heat and mechanical energy, the first news of which was given at the Cork meeting of the British Association in 1843[26]. The name was given to the unit in 1876. A full historical account of its history is given by Powell[27].

Brix
A specific gravity unit. (*See* **Degree (hydrometry).**)

Bushel
A unit of capacity equal to 8 gallons. The imperial bushel was legalized in Great Britain in 1826. The unit was originally defined in Magna Carta (1216) as the volume occupied by a quarter (500lb) of water; hence in those days, one gallon of water was taken to be $62\frac{1}{2}$lb.

Byte
The minimum amount of storage required by a computer to store an alphanumeric character or other special characters (e.g. space, question mark, plus sign etc.). Each character has a unique combination of 0 and 1 bits making 8 bits in all to make up its byte thereby enabling data to be recognized. The byte is sometimes called an octet.

C

Cable
The cable is a unit of length used at sea. It is not precisely defined[1] but its two most common values are 608 feet (1/10 nautical mile) and 720 feet (120 fathoms). The cable was in use by the middle of the sixteenth century.

Callier coefficient
A photographic plate appears to be more dense when viewed in parallel light than when it is examined in diffuse light. The ratio of the two densities is known as the Callier coefficient; it is equal to approximately 1·5.

Calorie (cal)
A unit of heat in the CGS system which was in use between 1880 and 1950. It represented the quantity of heat required to raise one gram of water through one degree Celsius (formerly Centigrade). Unfortunately the energy represented by a calorie varies according to the temperature of the water, thus there was the international steam calorie (4·1868 joules), the 15°C calorie (4·1855 joules), the 4°C calorie (4·2045 joules), and the mean 0–100°C calorie (4·1897 joules). These differences were overcome when the calorie was replaced by the joule as the primary unit of heat in the metric system in 1950[2], a step which was proposed by Lodge in 1895[3]. A large calorie sometimes called the kilocalorie or kilogram calorie is equal to 1000 calories. Physiologists in discussing metabolism use the kilocalorie as a unit, but they call it a calorie. (*See* **Kilocalorie**).

In France the calorie is often called the petit calorie and the grand calorie is equal to 10^3 petits calories. Chemists still use the calorie as a unit in thermo-chemistry, in which case one thermo-chemical or defined calorie is taken as equal to 4·1840 absolute joules.

The word calorie originated in France in 1787 when it was used to represent the 'matter of heat'. The word 'calorimeter' was used by Lavoisier in 1789[4]. The calorie received international recognition as the unit of heat in 1896[5], but it had been in use since 1880[6].

23

Candela (cd)

The candela is the SI base unit of luminous intensity. It is defined as the luminous intensity, in a given direction, of a source which emits monochromatic radiation of frequency 540×10^{-12} Hz (corresponding to a wavelength of about 550 nm) and whose radiant intensity in that direction is 1/683 watt per steradian. This definition was adopted[7] by the 16th CGPM (1980) when the definition promulgated[8] by the 9th CGPM (1948) was abrogated. The 1948 candela, sometimes called the new candle, had been defined in terms of the light emitted from a black body radiator at the temperature of freezing platinum, a method which had been proposed by Violle[9] in 1884 and had been put forward as a unit by Waidner in 1909[10] and ultimately adopted as an international standard some 40 years later. The candela, as defined in 1948 and amended in 1980, has a luminous intensity 1·9% less than the international candle which it replaced. The unit has at times been called the violle and came into use in Britain in 1948[11] and in the USA in 1950.

Candle, International

This is the unit of luminous intensity (candle power) which most countries agreed to use in 1909[12]. Germany, however, continued to use the Hefner candle (Hefnerkerze HK) unit which was derived from the Hefner lamp and had a luminous intensity of about 0·9 international candles. The old French unit of luminous intensity, the carcel, was equal to between 9·4 and 10 international candles. The bougie-decimale, another French unit, was equal to 0·104 carcel, and the candle used in the USA was 1·6 per cent larger than the international candle.

The candle was first legally defined in Great Britain in the Metropolitan Gas Act of 1860[13] and received recognition as an international standard at the International Electrotechnical Conference in 1881. The candle was specified as a spermaceti candle weighing six to the pound and burning at a rate of 120 grains in the hour. The definition mentioned neither the size of the wick nor the composition of the air, both of which control the luminous intensity of the flame. It was replaced by the candela (q.v.) in 1948.

Candle power

Candle power is the light radiating capacity of a source expressed in terms of a standard source of luminous intensity of one candela. Previous to 1948 the luminous intensity of the standard source was taken as one international candle, and therefore the candle power rating of all sources was lower by 1·9 per cent compared with the values in use today. (**Candela** q.v.).

Candle power was first legally defined in the Metropolitan Gas Act of 1860[13].

Carat (CM)

A unit of mass equal to 200 milligrams used for weighing precious stones[14]. The name is derived from quirrat, the Arabic name for the seeds of the coral tree, which were the traditional weights for precious stones. The 200 mg carat was introduced in Britain in 1913[15] having been approved six years before by the 4th CGPM. Previous to this the carat weighed about 205 mg, but its value altered slightly from country to country even though all reputable jewellers had agreed to use the 205 mg carat in 1877. The unit came into use in the sixteenth century.

The word carat is also used to indicate the purity of gold. Absolute purity is taken as 24 carat; an alloy of 18 parts gold and 6 parts impurity would be 18 carat. In the USA the purity of gold is spelt karat, but the mass of precious stones is expressed in carats.

The mass of a diamond is sometimes measured in points; one point is equal to 0·01 carat.

Carcel

An obsolete unit of luminous intensity used at one time in France[16]. 1 carcel≃10 international candles≃10 bougie-decimale.

Cascade unit

A unit of length used in cosmic ray work. (*See* **Shower unit**.)

Cé

A unit of time suggested at the Congrés International de Chronometer in 1900[17]. There were to have been 100 cé in a 'twenty four hour day'. The cé was divided into dedicé (1/10) and millicé (1/1000) to give periods of time analogous to the minute and the second.

Celo

A name suggested for the unit of acceleration in the British system of units. It is equal to one foot per second per second. The name is seldom used, one of its few appearances in print being in Latimer Clark's *Dictionary* in 1891[18].

Celsius scale

See **Temperature scales**.

Cent

A unit of musical interval. (*See* **Musical scales**.)

Cental (cH)

The cental, sometimes called the new hundredweight, was introduced in the Liverpool Corn Market on 1 February 1859 in an attempt to get uniformity in corn measures. It is equal to 100 pounds or one US short hundredweight and was legalized by an Order in Council dated 4 February 1879. Alternative names for the unit are the centner and the kintal.

Centihg
An abbreviation sometimes used for describing pressure in centimetres of mercury. Millihg is the corresponding term for pressure in millimetres of mercury. Both are pronounced with the *h* silent. The use of these abbreviated terms is deprecated by at least one writer[19].

Centigrade heat unit (Chu)
The centigrade heat unit is the energy required to raise the temperature of one pound of water through one degree Celsius (formerly Centigrade). One centigrade heat unit is equivalent to 455 calories, 1·8 Btu or 1900·4 J. The unit is often called a Chu, a word formed by its initial letters. The Chu is a hybrid between the British and metric systems and has never received an international standing. At the turn of the century it was comparatively popular in books on heat engineering.

Centimetre (electromagnetic)
An electromagnetic unit of inductance. A circuit has an inductance of one centimetre when an e.m.f. of one e.m. unit of potential is induced in it by a current changing at a steady rate of one e.m. unit of current per second. An inductance of 10^9 cm is called a henry (q.v.).

Centimetre (electrostatic)
An electrostatic unit of capacitance. A charged body has a capacitance of one centimetre if its potential is raised by one e.s. unit of potential when it is given a charge of one e.s. unit of charge. A capacitance of 900×10^9 cm is called a farad (q.v.).

Centimetre (length) (cm)
The CGS unit of length is the centimetre which is equal to 10^{-2} metre (q.v.).

Centner
An alternative name for the cental (q.v.).

Centrad
A unit of angular measure equal to 0·01 radian. It is used to specify the angular deviation of a beam of light by a narrow prism.

Century
A group of 100 things, e.g. 100 years.

Cetane/octane numbers

These are used in automobile engineering to describe the 'anti-knock' properties of compression ignition (diesel) and of spark ignition engine fuels. Both are based on the percentage of 'anti-knock' component in a standard fuel.

The cetane number is the percentage by volume of cetane added to methylnaphthalene to give the mixture the same 'anti-knock' properties as the diesel fuel under test. It was suggested as a standard in 1932 and approved by the ASTM (American Society for Testing Materials) in 1934[20]. High speed diesel engines require fuel with a cetane number above 40 for efficient operation.

The octane number is equal to the percentage by volume of trimethylpentane (*iso*-octane) which must be blended with *n* heptane to give the mixture the same 'anti-knock' characteristics as the petrol (gasolene) under test. it was proposed in 1927[21] and was defined by the ASTM in 1934[22]. In 1956 a suggestion was made and approved by the ASTM to extend the scale to octane numbers above 100 by adding tetra-ethyl-lead to the *iso*-octane[23]. High compression spark ignition engines need petrol with an octane number of 85 or more for efficient performance.

Chad

The chad is the name proposed in 1960[24] to represent a neutron flux of one neutron per square centimetre per second. The name is derived from that of Sir James Chadwick (1891–1974), a pioneer in nuclear physics. Some people[25] consider that if the chad be taken as 10^{16} neutrons per square metre per second, more convenient numbers would be obtained for the neutron fluxes usually met with in practice. This would make the flux of a power reactor, which is of the order of 10^{16} neutron $m^{-2} s^{-1}$ equal to one chad, and that of a sub-critical assembly (10^{10} neutron $m^{-2} s^{-1}$) equivalent to one micro-chad.

Chain

A Gunter's chain[26] of 22 yards length consists of a hundred links, so that each link is almost 8 inches in length. It was developed by Edmund Gunter in 1620 from the four perch line devised by Valentyne Leigh in 1577. A hundred foot chain is sometimes used in the USA. This is called an Engineer's or Ramsden's chain.

Chill wind factor

This is an arbitrary figure used originally by the Canadian Army in an attempt to correlate the performance of equipment and personnel with the conditions experienced in an Arctic winter[27]. The factor is obtained by adding temperature and wind velocity. The former is expressed as degrees below zero on the Fahrenheit scale and the latter in miles per hour. Thus conditions at $-25°$F with a wind of 15 miles h^{-1} have a chill wind factor of 40.

Chroma scale
A colour scale. (*See* **Surface colour.**)

Chronon
The shortest unit of time is considered by some to be time taken for electromagnetic radiation to cover a distance equal to the radius of an electron; this comes to about 10^{-23} second and it is called the chronon or tempon.

Circular mil
A unit of area used for describing the area of a conductor of circular cross-section. A conductor of diameter M mils has a cross-sectional area of M^2 circular mils. Hence π is not required in calculating an area of a circle if its diameter be in mils and its area in circular mils. The unit came into use about 1890[18].

Clark degree
A unit of water hardness (q.v.).

Clarke scale
A scale used in hydrometry. (*See* **Degree (hydrometry).**)

Clausius
A unit of entropy defined as Q/θ, where Q is the energy in kilocalories and θ the absolute temperature. The unit is used mainly by engineers. It is named after R. J. L. Clausius (1822–88) and came into use just before the 1939–45 war[28]. The unit is sometimes called the rank after W. J. M. Ränkine (1820–72), one of the originators of the thermodynamic temperature scale.

Cleanliness unit
The unit of measurement is expressed as the number of particles of a size of 0·5 micrometre or greater per cubic foot of air. A typical factory may have as many as a million of such particles per cubic foot (Class 10^6). A clean factory may have 100 000 particles (Class 10^5). Today, cleanliness is usually expressed in parts of contaminate by volume per million parts of air or in milligrams of contaminate per cubic metre of air (mg m^{-3}). Tables showing the maximum amount of contaminate which can be safely permitted in factories and other work places have been published by many of the industrial nations[29]. In these there is general agreement as to the limit

(threshold limit value – TLV) up to which operatives may be exposed for 40 hours per week with no incurred adverse effect and also on the maximum concentration to which exposure can take place for well spaced out periods of 15 minutes without serious risk to health. The former are described as TLV-TWA (time weighted average) and the latter as TLV-STEL (short-term exposure limit) dosages. In addition TLV-C tables are issued that give the concentrations which, if exceeded, could cause very serious health hazards. In general, contaminates are divided into four groups which are arranged in order of toxicity; the maximum allowed concentrations for each group are given in Table 2.

TABLE 2. *Permitted maxima of contaminates in air*

Group	Concentration ($\mu g\ m^{-3}$)	Typical substance
I	0–50	Beryllium; blue asbestos
II	50–250	silica
III	250–1000	coal dust
IV	1000 +	cement

Substances in Group IV are often of greater nuisance value than hazard to health.

Clo

The clo is a unit used in the measurement of the thermal insulation of clothing[30]. It is the amount of insulation necessary to maintain comfort and a mean skin temperature of 92°F in a room at 70°F with air movement not over 10 feet per minute, humidity not over 50% with a metabolism of 50 kilocalories per square metre per hour. Assuming 76% of the heat is lost through the clothing a clo is defined in physical terms as the amount of insulation that will allow the passage of 1 calorie $m^{-2}\ h^{-1}$ with a temperature gradient of 0·18°C between the two surfaces

So 1 clo $= 0·18°C\ h\ m^2\ cal^{-1}$
$= 0·648°\ C\ s\ m^2\ cal^{-1}$
$= 0·154°C\ m^2\ W^{-1}$
$= 0·88°F\ h\ ft^2\ Btu^{-1}$

It was originally defined in 1941 as the insulation value of a business suit with the usual undergarments which maintain a sedentary man at a comfortable temperature under normal indoor conditions as found in the USA. The best clothing has a value of about 4 clo per inch of thickness. The theoretical value for absolutely still air at 18°C is estimated as 7 clo.

Clusec
The power of evacuation of a vacuum pump is sometimes expressed in clusecs, 1 clusec being defined as 1 centilitre per second at a pressure of 1 millitorr. Thus the clusec has the dimensions of power and is approximately equal to 1.3×10^{-6} W.

Comfort index
This is used as an arbitrary guide to acquaint travellers of the working conditions (climatic) in the warmer regions of the world[31]. Both temperature and relative humidity (expressed as a percentage) are involved, the index being obtained by adding the local temperature (°F) and (relative humidity)/4; when the total is less than 95, working conditions can be expected to be tolerable; for values between 96 and 103 one can work provided one has been conditioned to the tropics where a temperature of 91°F with 40% relative humidity is no more uncomfortable than one where it is eight degrees cooler but with a humidity of 80%. In both instances it is assumed there is no breeze, the presence of which would enable higher temperatures to be endured.

Compton wavelength
See Appendix 1.

Condensation number (Co)
A name given to the ratio of the number of molecules condensing on a surface to the total number of molecules striking the surface. It is used in convection.

Cord
A unit used to measure volume of timber. 1 cord = 1536 board feet = 128 ft^3 = 3.62456 m^3.

Coulomb (C)
The practical and the SI unit of charge. It is the quantity of electricity transported in one second by a current of one ampere. From 1908 to 1948 the international coulomb, derived from the international ampere, was in use. Like the other international units it was replaced by the absolute unit on 1 January 1948[32].

The name coulomb was given to the unit at the first meeting of the IEC in Paris in 1881[33]. At this meeting two of the five units, which were given definitions, were named after French scientists. These were the ampere (A. M. Ampère 1775–1836) and the coulomb (C. A. Coulomb 1736–1806).

(1 international coulomb = 0.99985 absolute coulomb).

Cowling number (Co)
A dimensionless number used in magnetohydrodynamics given by $B^2/\mu\rho v^2$

where B is magnetic flux density, μ is permeability, ρ is density and v is fluid velocity[34]. It is named after Professor T. G. Cowling, who was one of the pioneers in magnetohydrodynamics.

Cran

A unit in the herring fishing industry used for indicating quantity of fish. One cran of herrings occupy a volume of 37·5 imperial gallons, so that a cran represents about 750 fish. The measure was in use at the end of the eighteenth century, and was originally the quantity of fish needed to fill a barrel. The cran has been mentioned in several Acts of Parliament over the period 1815–1908[35].

Crinal

The unit of force in the decimetre kilogram second system[36]. One crinal is equal to 10^{-1} N. The name is derived from crinis meaning a hair, because it was considered that a force of one crinal (about 10 grams weight) might just break a hair. The decimetre kilogram second units were proposed in 1876 but were never adopted. The name is attributed to James Thompson, the elder brother of Lord Kelvin.

Crith

A unit of mass used for weighing gases. It is equal to the mass of one litre of hydrogen at standard temperature and pressure ($90·6 \times 10^{-6}$ kg). The unit was proposed by Hofmann[37] in 1865 who derived its name from the Greek for a barley corn. The mass of a litre of any gas at s.t.p., when expressed in criths, is numerically equal to half its molecular weight, e.g. oxygen has a mass of $16/2 = 8$ criths.

Crocodile

A name sometimes used in certain nuclear laboratories in Britain to express electrical potentials of a million volts.

Cron

A name suggested by J. S. Huxley in 1957[38] for a unit of time equal to a million years; thus a millennium is equal to a millicron. The word is derived from the Greek word for time.

Cubit

This is the earliest unit of length and it originated in Egypt in the Third Dynasty (2800–2300 B.C.). Other forms of cubit were adopted elsewhere in the Middle East and in Mexico. All these cubits were about 20·6 inches long so it is probable they were derived from the same source which is believed to be the human fore-arm.

Cumec

A unit of liquid flow in which 1 cumec represents a flow of 1 cubic metre per second.

Curie (Ci)

A unit of radioactivity which is now defined as the quantity of any radioactive nuclide in which the number of disintegrations per second is $37{\cdot}00 \times 10^9$. The unit was adopted at a Radiography Conference in Brussels[39] in 1910 when it was defined as the radioactivity associated with the quantity of radon in equilibrium with on gram of radium. The present definition, which refers to a unit of the same size but described in terms independent of the disintegration of radon, was agreed at the Copenhagen meeting of the International Commission on Radiological Units[40] in July 1953. The unit is named after Pierre Curie (1859–1906), one of the discoverers of radium. The curie is too large for normal laboratory work where the radioactivity is generally of the order of millicuries.

Curie scale

See **Temperature scales**.

Cusec

A unit giving the speed of pumping of a pump. It is equal to one *cu*bic foot per *sec*ond but the pressure is not specified[41]. The name is of pre 1939–45 war origin.

D

D unit

A unit of X-ray dosage introduced by Mallet[1] in 1925. One D unit was equal to 102 roentgen. It is now obsolete.

Dalton

The name occasionally used for the atomic mass unit (q.v.). This unit is named after John Dalton (1766–1844), the instigator of the atomic theory of matter.

Daraf

The unit of elastance or reciprocal of capacitance. The name is derived by writing farad backwards. The unit was used by Kennelly[2] in 1936 but few other writers employ it.

Darcy

The permeability of a porous material is measured in darcys. This CGS unit was proposed in 1933 and is named after H. Darcy (1803–58)[3], the French scientist who investigated the flow of fluids in porous media in 1856. The darcy is defined as the volume of liquid of unit viscosity (1 centipoise) passing through unit area of a porous medium in unit time when subjected to a pressure gradient of 1 atmosphere per unit distance. The permeability of sand filters varies from about 25 to 140 darcys; that of sandstone is of the order of one or two darcys.

Darwin

The darwin is the name suggested for a unit of evolutionary rate of change. When a species increases or decreases by an exponential factor e in a million years, its rate of change is one darwin; this rate is also equivalent to a change of 10^{-3} in a thousand years. Rates of change in nature seldom exceed one darwin, but with domestic animals changes of the order of a kilodarwin are often found. The unit was proposed by J. B. S. Haldane in 1948[4]; it is named after Charles Darwin (1809–82) the controversial Victorian biologist who put forward the theory of evolution.

Day (d)

The sidereal day is the mean time taken for the earth to complete one revolution. The solar day is the mean time interval which elapses between when the sun is in a predetermined position, e.g. overhead, and its return to the same position. There are 365·24 solar days and 366·24 sidereal days in a year. There are 86 400 seconds in a mean solar day and 86 164·0906 seconds in a sidereal day.

Decade

A group of ten, e.g. ten years.

Debye unit (D)

A unit of electric dipole moment equal to 10^{-18} e.s.u. cm or $3·336 \times 10^{-30}$ coulomb metre. The unit[5] evolved naturally from the fact that many electric dipole moments for molecules are a multiple of 10^{-18} e.s.u. cm. Thus the dipole moment of hydrochloric acid is $1·05 \times 10^{-18}$ e.s.u. cm, for aniline it is $1·5 \times 10^{-18}$ e.s.u. cm. The unit first received a name in 1934[6], but it is frequently used in the abbreviated form, which is the letter D. It has also been defined as the product of the electronic charge and either the radius of the first Bohr orbit of hydrogen giving a value of $2·54 \times 10^{-18}$ CGS units or one ångström giving a value of $4·803 \times 10^{-18}$ CGS units. It is named after P. J. W. Debye (1884–1966) the pioneer authority on polar molecules.

Decibel (dB)

A unit which expresses the rates of either power, intensity levels or pressure levels and which is used mainly in acoustics and telecommunications. The powers P_1 and P_2 of two sources differ by N decibels where $N = 10 \log_{10} P_1/P_2$—that is $P_1/P_2 = 10^{0·1N}$. A wave of intensity I has an intensity level of N decibels when referred to another wave of intensity I_0 where $N = 10 \log_{10} I/I_0$. When dealing with audible sounds the middle frequency is about 10^3Hz and I_0 is taken to be the threshold of audibility which for sound waves in air is 10^{-12} $W\,m^{-2}$. When dealing with sound pressures $N = 20 \log_{10} P/P_0$ where P and P_0 are the r.m.s. sound pressures of the two sources. N is called the sound pressure level when P_0 is the reference sound pressure of 2×10^{-5} $N\,m^{-2} = 2 \times 10^{-4}$ μbar. Loudness levels are expressed in phons (q.v.) and are measured by sound level meters which measure certain weighted sound pressure levels. The weighting as a function of frequency is given by three curves A, B and C. Curve A gives the best agreement with the subjective sensation of sound and so sound pressure levels are expressed in dB(A). Loudness levels do not correspond to the sensation of loudness which is measured in sones (q.v.) and which cannot be measured directly but can be calculated from loudness levels. The decibel and the phon were the only units defined at the first International Acoustic Congress which was held in Paris in 1937[7].

The following names have been suggested in recent years for the decibel.

The logit[8] was proposed in 1952, the decilit[9] (from *decilogarithmic unit*) in 1955 and the decilog[10], decomlog[11] and decilu[12] appeared in 1954. The decilit was a product of the Bell Telephone Laboratories.

Decilit

A name suggested in 1955 for the decibel (q.v.).

Decilog (pressure)

A unit of pressure. (*See* **Logarithmic scales of pressure**.)

Decilog (ratio)

A name suggested in 1954 for the decibel (q.v.).

Decilu

A name suggested in 1954 for the decibel (q.v.).

Decomlog

A name suggested in 1954 for the decibel (q.v.).

Decontamination factor (DF)

The decontamination factor is used in radiological protection; it is the ratio of the original contamination to the residual contamination after decontamination[13]. The ease of decontamination (ED) is a qualitative term which is described as excellent for values of DF above 1000, good (100–1000), fair (10–100) and poor for ratios of less than 10.

Degré

A unit of time discussed at an international conference on time measurement held in 1900[14]. It was suggested the day be divided into 100 parts each to be called a degré. The degré would be subdivided into grades, namely decigrade, centigrade, milligrade and decimilligrade. It was proposed that the popular names for these divisions would be minute premiere, minute second, moment and instant respectively.

Degree

The term degree has been used since the fourteenth century to indicate the quantity of certain physical quantities such as angles and temperatures. In the last two centuries the use of degrees has increased to include units of hydrometry, hardness, viscosity and many others.

Degree (angle)

A unit of angular measure in which a quadrant of a circle is divided into 90 equal parts. This subdivision of the circle into 360 degrees can be traced back to Babylonian times.

Degrees of latitude have been measured from an arbitrary zero at the terrestrial equator since the beginning of the sixteenth century, The zero for degrees of longitude varied from country to country until it was agreed at the Meridian Conference held at Washington in 1884[15] that longitude be measured from an arbitrary zero at Greenwich. Several countries did not subscribe to this convention until the twentieth century. One of the last to adopt it was France who had previously used the Isle de Fer ($18°$W, $27°$ $45'$N) as the zero of their system.

Degree (API gravity)

This is used in the petroleum industry to describe the density of petroleum products. A degree (API) is given by the number ($141·5$/(specific gravity at $60°$F)$-131·5$). Values lie within the range -1 to $+101$, the larger API numbers referring to lighter oils and vice versa. The term API gravity (American Petroleum Institute) was authorized in the United States[16] in 1952 and is usually found by a specially calibrated direct-reading hydrometer.

Degree (hydrometry)

This degree refers to hydrometry readings on an arbitrary scale[17]. Equations are available to convert the readings into actual values of specific gravity. The degrees are the API, Bailling (1835), Bates (1918), Baumé (1784), Beck (1830), Brix (1854), Cartier (1800), Gay Lussac (1824), Sikes (1794) and Twaddell (1830). Bailling, Baumé (sometimes called Lunge) and Brix degrees are used when giving sugar content in aqueous solutions. Sikes degrees refer to the alcohol content of a solution. These degrees are now obsolete since in 1912 it was suggested that all previously named hydrometer degrees should be replaced by a universal hydrometer degree defined as 100 s, where s is the specific gravity of the liquid concerned.

According to Glazebrook writing in the *Dictionary of Applied Physics* the scales adopted for hydrometers are infinite in variety. Some of these are the Clarke (1730), Dycas (1790), and Francoeur (1842) for estimating alcohol content of a liquid and the Quevenne, Soxhlet and Vieth scales used in lactometry. Correlation tables for the more common hydrometers are given by Preston[18].

Degree (photographic)

The Scheiner and DIN systems for indicating photographic emulsion speed are sometimes called degrees. (*See* **Photographic emulsion speed indicators**.)

Degree (viscometry) (Engler, MacMichael)

The Engler degree is used in viscometry to indicate the ratio of the times required for 200×10^{-6} m^3 of liquid and for the same volume of water at $20°$C to pass through an Engler viscometer[19]. Conversion tables are available (Erk, *Forschumps-arbeiten* VDI No. 288 (1927)) to relate Engler degrees to the flow times obtained with other viscometers. Thus for example one Engler degree is equivalent to a dynamic viscosity of 10^{-5} m^2 s^{-1}, to $51·7$ seconds with a Redwood No. 1 viscometer at $70°$F and to $58·18$ seconds with a Saybold Universal viscometer at $100°$F. The Engler degree is of

German origin, the Engler viscometer being the first efflux viscometer and it was in use as early as 1884[20].

In the MacMichael viscometer the viscosity of a liquid is indicated in terms of the torque exerted on an inner cylinder or disc which is in a rotating liquid. The torque is measured by the twist of the suspension of the inner cylinder which is given on an arbitrary scale and the viscosity is expressed as so many divisions, i.e. MacMichael degrees, on this scale. The relationship between dynamic viscosity of a liquid expressed in poises and the viscosity in MacMichael degrees depends upon the stiffness of the suspension of the inner cylinder and hence the gauge of the suspension wire is generally quoted along with the viscosity in MacMichael degrees. Thus for example 200 poises are equivalent to 330 MacMichael degrees with a 26 SWG suspension and to 560 MacMichael degrees with a 30 SWG. The MacMichael degree was introduced in 1915[21].

Degree (water hardness)
A unit expressing the hardness of water. (*See* **Hardness (water).**)

Demal (D)
A demal solution is one which contains 1 gram equivalent of solute per cubic decimetre of solution. If the volume should be taken as a litre the solution would be weaker by a factor of $1/1\cdot000028$. The name demal was proposed by H. C. and E. W. Parker in 1924[22].

Denier
A unit used in the textile industry to indicate the mass in grams per 9000 metres of yarn. (*See* **Yarn counts.**)

Dex
A name suggested for any ratio when expressed as a logarithm to the base ten, thus a ratio of 10^{29} would be 29 dex, and the ratio corresponding to an octave would be 0·301 dex. The name is derived from the initial letters of the words *d*ecimal *ex*ponent. It was originally proposed in 1951 and revived by J. B. S. Haldane in a letter to *Nature* in 1961.

Dioptre
The metric unit in which the power of a lens is measured. It is equal to the reciprocal of the focal length when the latter is measured in metres, so that a lens with a focal length of 0.5 m has a power of 2 dioptres. In opthalmic work the powers of converging lenses are considered to be positive, those of diverging lenses negative.

The dioptre was adopted as a unit at a medical conference in Brussels[23] in 1875 but the idea of defining the power of a lens as the reciprocal of its focal length was suggested seven years earlier by Nagel[24].

Dollar
A unit of reactivity which was sometimes used in the USA to describe the degree of departure of a nuclear reactor from its critical condition. It was first suggested in the 1940s.

Donkey power
A name suggested for a unit of power equal to 250 watts. The unit was proposed facetiously in 1884[25], and was chosen so that three donkey power was approximately equal to one horse power.

Dram or Drachm
A unit of mass used in the apothecaries' and the avoirdupois systems. In the former it is equal to $\frac{1}{8}$ ounce and in the latter to $\frac{1}{16}$ ounce. The name is derived from the Greek drachma and the unit itself was in use by the fifteenth century.

Drex
A unit used in the textile industry in Canada and the USA to indicate the mass in grams per 10 000 m of yarn. (*See* **Yarn counts**.)

Duffieux
A name suggested in 1973 for an SI unit of angular spatial frequency. It is named after P. M. Duffieux (1891–   ).

Duty
The name used extensively to describe work in the first half of the nineteenth century. It was equal to the work done in lifting a pound weight through a distance of 1 foot. The unit was introduced by James Watt[27] (1736–1819) and was exceedingly popular with engineers[28]. (*See* Foot pound.)

Dycas scale
A scale used in hydrometry. (*See* **Degree (hydrometry)**.)

Dynamode
A unit of work proposed in 1830 by G. C. Coriolis (1792–1843), the French physicist who gave an explanation for the forces which govern the movement of winds and currents in the oceans and set out the basic formula for kinetic energy: $\frac{1}{2}mv^2$. The unit indicated the work done in raising 1000 kg through one metre but it never came into general use.

Dyne (dyn)
The CGS unit of force. It is the force which will accelerate 1 gram by 1 centimetre per second per second, so that a gram weight represents a force of about 980 dynes. The unit was adopted by the British Association[29] in 1873. Professor Everett of Belfast is reputed to have suggested the title, the name being derived from the greek word to push.

E

e unit
A unit of X-ray dosage used by W. Friedrich in 1916[1]. One e unit represented a dosage which varied between 6 and 8 roentgens. It is now obsolete.

E unit
A unit of X-ray intensity used by W. Duane in 1914[1]. One E unit was equivalent to about one roentgen per second. It is now obsolete.

Earthquake scales
The intensity of an earthquake may be described by the Mercalli scale. This is a numerical scale in which intensities of increasing magnitudes are indicated by the Roman numbers I to XII. Thus a feeble earth tremor detectable only by a seismograph would be classified as I whereas a class VII earthquake would be sufficiently violent to ring church bells. In a grade XI earthquake railway tracks would be distorted; the material destruction in a XII class earthquake is not precisely defined. The scale was proposed by G. Mercalli in 1902 who adopted it from an earlier one introduced by M. S. Rossi and F. A. Forrell in 1883. The original Mercalli was modified in 1931; the new version is called the Modified Mercalli or MM scale[2].

The scale now generally used to give the absolute intensity rating of an earthquake is the Richter[2] scale. This scale reading 1, 2, 3, etc. was put forward in 1932 by Dr Charles Richter (1900–1985) while working at the Seismological Laboratory of the Californian Institute of Technology. It was initially used to distinguish between large and small earthquakes in California. The scale is based on the amplitude of the trace of a specified seismograph and a law devised by Richter on the attenuation of energy which enables the intensity of the tremor at the epicentre to be calculated. It was extended with the co-operation of Professor Beno Gutenberg to catalogue and classify earthquakes all over the world and has now been in general use since 1935. On the Richter scale, each point represents an intensity ten times greater than the one below it. There would appear to be no upper limit but in practice there has not been a reading above 9.

Einstein unit

A unit used to describe the photoenergy involved in a gram molecule of a substance during a photochemical reaction. It is equal to *Nhv* where *N* is Avogadro's constant, *h* is Planck's constant and *v* is the frequency of the electromagnetic radiation. The unit was in use in 1940[3] and is named after A. Einstein (1879–1955) who explained the photoelectric effect in 1905.

Ekman number

This number was used by the Swedish scientist V. W. Ekman in his work on ocean currents. It expresses the ratio of the viscosity force to the Coriolis force.

Electron volt (eV)

A unit employed to indicate the energy of a charged particle in terms of the energy received by the charge on an electron ($160 \cdot 218 \times 10^{-21}$ coulomb) due to a potential difference of 1 volt. It is equal to an energy of $160 \cdot 218 \times 10^{-21}$ joule. An approximate value (1 in 10^4) for the energy of electromagnetic radiation expressed in electron volts is given by $1234/\lambda$ where λ is the wavelength in nanometres.

In recent years it has become customary to write MeV and GeV for mega (10^6) and giga (10^9) electron volts. In the USA 10^9 electron volts are often written as BeV, the letter B being used in this case as an abbreviation for the American billion (10^9), but in 1948[4] the International Union of Pure and Applied Physics disapproved of the use of BeV and expressed a preference for GeV or 10^9 eV.

The electron volt was called the equivalent volt when it was originally introduced in 1912[5].

Eman

A unit which was at one time used to describe the radioactivity content of an atmosphere. One eman was equivalent to 10^{-10} curie per litre.

The eman was one of the units discussed at the International Radium Standards Committee in 1930[6], when it was said the unit had been in use since 1921 but no references to any publications containing the eman are given by the committee. It seems probable the name was derived from the first four letters of *eman*ation. The unit is now obsolete. $3 \cdot 64$ emans equal a Mache unit (q.v.).

Engler degree

A unit used in viscometry. (*See* **Degree (viscometry)**.)

English degree
A unit of water hardness. (*See* **Hardness (water).**)

Enzyme unit (U)
One unit of any enzyme is defined as that amount which will catalyse the transformation of one micromole of substrate per minute, or, where more than one bond of each substrate is attacked, one microequivalent of the group concerned under defined conditions of temperature, substrate concentration and pH number. Where two identical molecules react together the unit will be the amount which catalyses the transformation of two molecules per minute. Using this definition the concentration of an enzyme will be expressed as activity and will be given as units per milligram[7]. It may also be expressed as specific activity and given as units per milligram of protein.
   The unit is also called the International Union of Biochemistry unit.

Eon
A unit of time equal to 10^9 years which was suggested in 1968[8].

Eotvos unit (E)
A unit used in geophysical prospecting to indicate the change in the intensity of gravity with change in horizontal distance. It has the dimensions of 10^{-9} gal per horizontal centimetre. Changes in the earth's gravitational field are generally within the range 5 to 50 E. The unit was proposed by Barton in 1929[9] and is named after Baron Roland von Eotvos (1848–1919), the Hungarian physicist who, early in the twentieth century, made the first successful torsion balance.

Equal listener response scale (ELR)
An arbitrary noise scale used in acoustics. (*See* **Subjective sound ratings**.)

Equi-viscous temperature (EVT)
A unit for viscosity used in the tar industry. It is defined as that temperature in degrees Celsius at which tar has a viscosity of 50 seconds when measured in a standard tar efflux viscometer, e.g. 94 EVT is equivalent to a kinematic viscosity of about 200 stokes at 100°C. The unit was first suggested by G. H. Fuidge, of the South Metropolitan Gas Company about 1930. It is recognized by the British Standards Institution[10] and is in general use throughout Europe.

Erg
The CGS unit of work, i.e. the work done by a force of 1 dyne acting over a distance of 1 centimetre. The unit was proposed by the British Association in

1873[11] and the name was suggested by Professor Everett[12] from Queen's College, Belfast. It is derived from the Greek verb to work.

Ergon

A name suggested by Partington in 1913[13] for the product of Planck's constant and the frequency of the electromagnetic radiation concerned. The name ergon was derived from that used by the translator of a paper by Clausius[14] in 1868 to represent the unit of work. (*See* **Quantum.**)

Erlang (E)

A telephone traffic unit defined by the relationship $e = CT$, where e is the traffic density in erlang, C is the number of calls per hour, and T is the average time of each call. The unit was adopted in 1946 and is named after A. K. Erlang (1879–1924), a Danish mathematician who was interested in circuit utilization[15]. The erlang is sometimes called a traffic unit.

Euler number (Eu)

A number used in fluid dynamics defined by $p/\rho v^2$, where p is pressure, ρ density and v velosity. It is named after the German mathematician L. Euler (1707–83).

f number

The light-gathering power of telescope and camera objective lenses is expressed by the relative aperture which, provided the object is at infinity, is the ratio of the focal length to the diameter of the entrance pupil. For a single lens the entrance pupil coincides with the lens itself. In photography the relative aperture is called the f number and is considered to indicate the speed of the lens[1].

Towards the end of the nineteenth century lens speeds in photography were sometimes marked in the so-called uniform system (US) in which the speed number was proportional to the exposure time. In the f system the f number is proportional to the square root of the exposure time. The latter system was introduced by the Royal Photographic Society in 1881 and approved by an International Conference in Paris in 1900. Two f systems are in use and in both each f number represents a 50 per cent reduction in speed compared with the number preceding it. The systems are given in Table 3, the first being used by English-speaking countries while the second is common in Europe. The US numbers are 1, 2, 4, 8, 16, 32, in which 1 corresponds to $f/4$.

TABLE 3. *Standard f numbers*

US number							1		2		4		8		16
f number (English)	1·4		2		2·8		4		5·6		8		11		16
f number (Continental)		1·6		2·3		3·2		4·5		6·3		9		12·5	

Fahrenheit scale
See **Temperature scales**.

Fanning friction factor (f)

A dimensionless number used in describing turbulent isothermal flow in pipes. Its value for smooth pipes in terms of Reynolds' number Re when this lies between 5000 and 200 000 is given by $f = 0{\cdot}046/(Re)^{0{\cdot}2}$.

Farad (F)

The practical and the SI unit of capacitance. A capacitor has a capacitance of 1 farad when a charge of 1 coulomb raises the potential between its plates to 1 volt. The farad is too large for most practical purposes and the capacitance of laboratory capacitors is of the order of microfarads (10^{-6} F) or picofarads (10^{-12} F): the latter are sometimes called 'puffs'.

The unit, named after Michael Faraday (1791–1867), was first suggested by Latimer Clark[2] in 1867. The Latimer Clark farad was the same value as the present microfarad and was equal to the capacitance of about a third of a nautical mile of submarine cable. At the first meeting of the International Electrotechnical Conference in 1881 the farad, as used today, was defined. Several years previous to this, however, some authors had already begun to call the Latimer Clark farad a microfarad[3]. In 1948[4], when the absolute units replaced the international units, the international farad was superseded by the absolute farad.

$$(1 \text{ international farad} = 0{\cdot}99951 \text{ absolute farad})$$

Faraday

The Faraday, or Faraday constant, was originally defined as the quantity of electricity which would deposit 1 kilogram equivalent of silver from a conducting solution. It is now given in terms of the electricity required to deposit 1 mole of a monovalent ion from a conducting solution. Its present value is given in Appendix 1. The unit is named after Michael Faraday (1791–1867) the inventor of the dynamo and one of the greatest practical scientists of the nineteenth century.

Fathom

A unit of length of 6 feet used for describing depth of sea and, in the nineteenth century, for measuring distances in mines. It is very old[5]; references are made to it by Sir Henry Mainwaring in his *Seaman's Dictionary* published in 1644 and by L. C. Wagenaer some sixty years earlier. It is claimed the name fathom is derived from the Anglo-Saxon faethm, to embrace, i.e. it is distance between the hands when the arms are held outstretched.

Fermi

A unit of distance equal to 10^{-15} m. It is used for describing nuclear distances just as the ångström is used for atomic distances. It is named after E. Fermi (1901–54) who built the first atomic pile; the unit was first used in 1956.

Fineness

The fineness of gold in an alloy is expressed as the parts of gold in 1000 parts of alloy. Thus pure gold has a fineness of 1000, and gold with 0·1 per cent of another metal has a fineness of 999.

Finsen unit (FU)

An ultraviolet ray of wavelength 296·7 nm has an intensity of 1 finsen unit when its energy density is 10^5 W m^{-2}. A ray of 2 FU will cause sunburn in 15 minutes. The unit is named after N. R. Finsen (1860–1904) an advocate of ultraviolet therapy[7].

Fire danger rating index

This is an arbitrary index which ranges in value between 0 and 100. It is used by the Forestry Commissioners in Great Britain to indicate the probability of the outbreak of fire in woodlands. The index depends upon the relative humidity of the air, air temperature, wind velocity and rainfall. An index of zero indicates the outbreak of fire is improbable, e.g. during prolonged rain in mid-winter. The figure 100 implies there is a serious fire hazard such as would be expected during a drought in a hot summer.

Flux unit (fu)

In radio astronomy energy is usually measured in units of 10^{-26} W m^{-2} Hz^{-1}. The factor seems to have been first used in 1951[8]; it is interesting to note that radio astronomers have always measured energy flux in terms of square metres. (*See* **Jansky**.)

Foaminess (Σ)

A CGS unit suggested in 1938[9] to indicate foaminess. It is defined by vt/V, where v is the volume of foam produced by passing V cm^3 of air through a liquid in t seconds. Thus for 1 per cent *n*-butyl alcohol the foaminess is 4·7 seconds. No name has been proposed for the unit, but it has been suggested that it be represented by the symbol Σ which is the initial letter of the Greek word meaning lather.

Foot

A unit of length equal to one third of a yard. The foot has been a unit of length in England for over a thousand years. In Mauritius and parts of Canada the French foot (12·8 inches) is sometimes used.

Foot-candle (fc)

A former unit of illumination equivalent to the illumination received at a distance of 1 foot from an international candle. It has been replaced by the lumen per square foot as the British unit of illumination.

$$1 \text{ foot-candle} = 10 \cdot 764 \text{ lux} = 1 \cdot 0764 \text{ milli-phot}.$$

Foot-lambert (ft L)

A unit describing the average luminance of any surface emitting or reflecting 1 lumen foot $^{-2}$, or expressing the luminance of a perfectly diffusing surface. A completely reflecting surface under an illumination of 1 foot-candle has an average luminance of 1 foot-lambert; the average luminance of any surface in foot-lamberts is the product of the illumination in foot-candles and the surface reflection factor.

$$1 \text{ foot-lambert} = 1 \cdot 076 \text{ milli-lamberts} = 1/144\pi \text{ candles inch}^{-2}.$$

This unit was adopted by the Illuminating Engineering Society of New York and by the American Engineering Standards Committee[10]. It is now obsolete[11].

Foot pound (ft lb)

The unit of work in the foot pound second system of units. It is the work done when a force of 1 pound weight (c 32 poundals) is applied over a distance of 1 foot. The unit was used by James Watt in the latter half of the eighteenth century. At first it was known as the duty, but from the middle of the nineteenth century it has been called the foot pound[12].

Fors (f)

A name suggested by the SUN Committee (International committee for the correlation of Scientific Symbols, Units and Nomenclature) in 1956 for the force represented by 1 gram weight[13]. At the same time this committee proposed that the force due to a kilogram weight should be called a kilofors – kf. 1 fors is equal to g dynes, where g is the numerical value of the acceleration due to gravity in CGS units. The name is derived from the Latin for force.

Fourier

A name proposed by Harper[14] in 1928 for the metric unit of thermal conductivity. It is equivalent to watts C^{-1} cm^{-1}. The unit is named after the French scientist J. B. J. Fourier (1768–1830) who defined thermal conductivity in 1822[15]. The name Fourier was considered for a second time in 1931[16], but was never used and has been discarded in favour of the thermal ohm (q.v.).

Fourier number (Fo)
A dimensionless number used in heat transmission. it is equal to

$$\frac{k}{c\rho}\frac{t}{l^2} \quad \text{or} \quad \frac{Kt}{l^2}$$

where k is the thermal conductivity, t represents time, c is specific heat, ρ is the density, l has the dimensions of length and K is the thermal diffusivity or thermometric conductivity. The number is named after J. B. J. Fourier (1768–1830) the French scientist who put the theory of heat on a firm mathematical basis[17].

The Fourier number (Fo*) used in mass transfer problems is given by Dt/l^2 where D is the diffusion coefficient.

Francoeur scale
A scale used in hydrometry. (*See* **Degree (hydrometry).**)

Franklin
A name suggested[18] in 1941 for unit charge in the electrostatic CGS system of units. It was adopted by the SUN Committee in 1961 who defined it as that charge which exerts, on an equal charge at a distance of 1 centimetre in vacuo, a force of 1 dyne. Hence 1 franklin $=(10/c)$ coulomb where c is the velocity of light in cm s^{-1}. The unit is named after Benjamin Franklin (1706–90), one of the early investigators of electrical phenomena.

Fraunhofer
This is used to describe spectral width as a function of wave length; it is defined as $(\delta\lambda/\lambda) \times 10^6$, where $\delta\lambda$ is the line width and λ its wave length. The unit is named after J. von Fraunhofer (1787–1826) who was one of the first people to give a satisfactory explanation for the dark lines in the solar spectrum.

French
A unit used to denote small diameters particularly in fibre optic bundles[19]. The diameters are expressed on the Charrière or French scale, each unit of which is equal to 1/3 mm.

French degree
A unit of water hardness. (*See* **Hardness (water).**)

French vibration
A French vibration may be defined by the relationship $T = \pi[I/M]^{1/2}$, where T is the periodic time, I and M are the moment of inertia and the mass

respectively of the moving system. Tuning forks marked with their frequency in French vibrations carry a number equal to $2f$, where f is the normal frequency expressed in cycles second^{-1}. The French vibration was legalized in France in 1859 when the French decided to have a standard pitch for their musical scales[20]. The standard corresponds to a frequency of 435 cycles second^{-1} but was defined as a diapason normal of 870. The French vibration is now obsolete but old tuning forks sometimes appear in laboratories with their frequencies marked in French vibrations instead of in cycles second^{-1}.

Fresnel
A unit of optical frequency which is sometimes used in spectroscopy. One fresnel is equal to 10^{12} Hz; it can be obtained by multiplying the wave number in kaysers by $10^{-10}c$, where c is the velocity of light in m sec^{-1}. Thus the frequency of a line of wavelength 500 nm can be expressed either as 20×10^3 kayser or $20c \times 10^{-7}$ fresnel. The unit came into use about 1930[21] but it has never been popular; it is named after A. J. Fresnel (1788–1827) an early exponent of the wave theory of light.

Frigorie
The frigorie is a unit of refrigeration used in continental Europe. It represents a rate of extraction of heat equal to 1 kilogram calorie per hour[22]. The unit is too small for modern commercial use and is now going into disuse.

Fringe value
A unit used in photo-elastic work. (*See* **Brewster**.)

Froude number (Fr)
A non-dimensional parameter used in fluid dynamics to describe the flow of a fluid in which there is a free surface. It is defined by the relationship $\mathrm{Fr} = v/(gl)^{1/2}$, where Fr is the Froude number, v is the velocity of the liquid, g the acceleration due to gravity and l is a parameter representing length. The flow of an inviscid, incompressible fluid in two geometrically similar flow systems is dynamically similar when their Froude numbers are the same. The number is named after William Froude (1810–79) who derived it in 1869[23] but is called the Reech number in France. It is used extensively by naval architects.

The number $1/\mathrm{Fr}^2$, which identifies the stability of flow in terms of velocity gradient at a fixed surface, is sometimes called the Richardson number (Ri) after Colonel A. R. Richardson of London University[24] who was very prominent in fluid mechanical research immediately before and after the 1914–18 war. Low values of Ri indicate the fluid is homogeneous.

Funal

A unit of force in the metre tonne second system of units; it is equal to 10^3 newtons. The name was proposed by James Thomson in 1876[25] and is derived from the Latin word meaning a rope. The unit is now called the sthéne (q.v.).

Furlong

A unit of length equal to 10 chains (220 yd); it dates from before the Norman conquest (1066) and is reputed to have been the length of the furrow ploughed along one side of the 10-acre common field of medieval England.

G

g scale

In the aircraft industry the force to which an aircraft may be subjected is often measured in terms of the force of gravity. Thus a force equal to six times that of gravity is called a force of 6 g.

G scale

A logarithmic scale of area proposed for geographical use in 1965[1]. If A is the area of the earth's surface ($198 \cdot 8 \times 10^6$ mile^2), then the G scale value of an area of A' mile^2 is given by $G = \log_{10}(A/A')$; thus a G value of 8 corresponds to an area of about two square miles.

G value

A unit used in radiation chemistry. It is the number of molecules destroyed or produced for each 100 eV of energy absorbed.

Galileo or gal

A unit of acceleration equal to 10^{-2} metre second^{-2}. It is used extensively in geological survey work where differences in the acceleration due to gravity of the order of milligals are frequently encountered. The name gal was adopted in Germany for the unit[2] about 1920 and is derived from Galileo Galilei (1564–1642), the Florentine scientist who transformed medieval natural philosophy into modern science. The name was most acceptable and was included in Glazebrook's *Dictionary of Applied Physics* in 1922[3].

Galileo (momentum)

The galileo was put forward in 1972 as a name for an SI unit of linear momentum[4], but the proposal has not been greeted with enthusiasm[5].

Gallon (gal)

A foot pound second unit of volume. The imperial gallon is the fundamental unit of capacity in the UK; it is defined by the Act[6] of 1963 as the space occupied by 10 pounds of distilled water of density 0·998859 gram millilitre^{-1} weighed in air of density 0·001217 gram millilitre^{-1} against weights of density 8·136 gram millilitre^{-1}. The gallon is mentioned in *Piers Plowman* (1342) by which time it had been in use as a legal standard for over 100 years. The imperial gallon is somewhat larger than the United States gallon which has its origin in the old English wine gallon of colonial days.

(1 imperial gallon = 1·20094 US gallon = 4·54596 litres

1 US gallon = 0·832674 imperial gallon)

Galvat
One of the names suggested at one time for the unit of current which today is known as the ampere. It is derived from L. Galvini (1737–98) of Bologna, one of the pioneers of electricity.

Gamma (magnetism)
A unit of magnetic intensity equal to 10^{-9} tesla. It has been used in geophysics since the beginning of the twentieth century and is employed mainly to describe annual changes in the earth's magnetic field. One of its first appearances in print[7] was in 1903 when it was defined as 10^{-5} CGS units.

Gamma γ (mass)
The microgram is sometimes called a gamma. In the International Critical Tables (1926) both names are applied to 10^{-6} g but in 1937 it was claimed that the gamma was being superseded by the microgram in the United Kingdom whereas the gamma reigned supreme in the USA and continental Europe[8].

Gamma γ (photography)
A unit employed in photography to give a measure of development. It is the slope of the Hurter and Driffield curve and measures the rate at which density grows as the exposure increases. The gamma of an emulsion depends on both the development time t and the developer, and as t is increased so the gamma rises to a limiting value called gamma infinity which is of the order of 1 to 3. The gamma was first used in 1903[9].

Gammil
A name suggested by Professor Conway of Dublin in 1946 for the unit of concentration in microchemistry[10]. A concentration of one gammil can represent either a microgram per millilitre, or parts per million or milligrams per litre. Alternative names suggested by Conway are the microgammil and the micril.

Gauss (Gs)

The gauss is the CGS unit of magnetic flux density. It is equal to a magnetic flux density of one maxwell per square centimetre, one line of magnetic induction per square centimetre, or 10^{-4} tesla. The gauss is comparatively small; the earth's magnetic field has a flux density of the order $2-3 \times 10^{-1}$ gauss.

The name gauss has been applied to several magnetic units. In Clark's dictionary[11] (1891) it is given as the unit of magnetic field strength and was equal to 10^8 CGS units. In 1895 the British Association[12] allocated the name to the CGS unit of magnetomotive force but five years later[13] the IEC adopted the gauss for the CGS unit of magnetizing force. In 1930 the unit for H was renamed the oersted[14] and the gauss became the name for the CGS unit of magnetic flux density, the unit it represents today. The name gauss is in honour of K. F. Gauss (1777–1855) an ardent advocate of the principle of expressing all physical quantities in terms of mass, length and time.

Gee pound

A name sometimes used to represent the mass which is accelerated by one foot per second per second by a force of one pound weight[15]. The gee pound is therefore equal to g pounds where g is the intensity of gravity in foot second $^{-2}$. It is also called a slug (q.v.).

Gemmho

A name sometimes used for the reciprocal megohm. (*See* **Mho**.)

Geological time scale

The duration and names of the geological periods which have occurred since precambrian times are given in the following table, all times being in millions of years.

Period	Began (million years ago)	Duration
Pleistocene	2	2
Pliocene	7	5
Miocene	26	19
Oligocene	38	12
Eocene	54	16
Palaeocene	65	11
Cretaceous	136	71
Jurassic	192	56
Triassic	225	33

Period	Began (million years ago)	Duration
Permian	280	55
Carboniferous	345	65
Devonian	395	50
Silurian	435	40
Ordovician	500	65
Cambrian	570	70
Precambrian	4600	4000

GeV

An abbreviation often used to express energy in units of 10^9 electron volts (q.v.).

Gibbs

A unit of adsorption suggested by Dean[16] in 1951. A surface concentration of 10^{-10} mole cm^{-2} is equal to one gibbs. The name commemorates J. Willard Gibbs (1839–1903) famous for his work on the phase rule.

Gilbert (Gb)

The gilbert is the unit of magnetomotive force which is equal to $1/4\pi$ abampere turns. There is no name, as yet, for the corresponding MKS unit which is one ampere turn. In 1895 the British Association[12], on the advice of Oliver Heaviside, adopted the name gauss for the CGS unit of magnetomotive force. Five years later[13] the IEC decided to use the name gauss for the unit of magnetizing force, thereby rendering the unit of magnetomotive force nameless. In practice, the name gilbert has been in use for the unit of m.m.f. since about 1903[17] but it was not internationally recognized until 1930[14].

The unit is named after William Gilbert (1544–1603) who is reputed to have made up the name electricity. The gilbert has never been widely used in practice. It appears in the forty-third edition of the *Handbook of Chemistry and Physics* (1961–62)[18] but has been omitted from the eleventh (1956) and subsequent editions of Kaye and Laby's tables[19].

Gill

A unit of volume equal to a quarter of a pint. The gill came into use as a measure for wine in the thirteenth century.

Glug

A name suggested for the mass which is accelerated by one centimetre per second by a force of one gram weight. It represents a mass of g grams, i.e.

approximately one kilogram and is the CGS counterpart of the slug. The name glug was proposed in 1957[20] but some writers claim the glug was used in some factories manufacturing grease before the 1939–45 war[21].

Gon
A gon is equal to a hundredth of a right angle[22]. An alternative name is the grade (q.v.).

Googol
The googol is the name for 10 to the power 100. The name is reputed to have been suggested by the nine-year-old nephew of E. Kasner (1878–1955) of Columbia University[23] when he was asked to name the largest number he could imagine. Subsequent prompting got the same source to produce the googolplex, which is 10 to the power googol.

Gouy
A name suggested in 1956[24] for an electrokinetic unit defined as $\zeta D/4\pi\eta$, where ζ is the electrokinetic potential, D is the electric displacement, and η is the electrolytic polarization. The unit has the dimensions of magnetic flux cm $^{-2}$, where the magnetic flux is given in CGS electrostatic units. It is named after L. G. Gouy (1854–1926) the French magnetic physicist.

Grade
A unit of angular measure in which a quadrant of a circle is divided into a hundred equal parts. It was one of the metric units sanctioned during the French Revolution. Today it is sometimes used in France but is seldom employed elsewhere. An angle of θ grades is written as θ^g.

Graetz number (Gz)
A dimensional number associated with the transfer of heat by streamline flow in a pipe. The Graetz number is given by $\pi d RP/4L$, where d and L are the diameter and length of the pipe respectively, R is Reynolds' number and P is Prandtl's number. The number is named after L. P. Graetz (1856–1941) the original investigator of heat transfer by modified laminar motion[25]. The number was recognized by the American Standards Association in 1941[26].

Grain (gr)
The smallest unit of mass in the English system. The grain is equal to 64.8×10^{-6} kg and is reputed to be named after a grain of wheat. The unit dates from the sixteenth century.

Gram (g)
A unit of mass in the CGS system[27]. The unit was at one time the

fundamental unit of mass in the metric system and was originally defined on 7 April 1795 as the mass of a thousandth part of a cubic decimetre of water. It was replaced by the kilogram as the fundamental unit within a few years of its introduction (*see* **Kilogram**). The name gram is derived from the Greek word meaning a small weight.

Gram-rad

A name sometimes used for the unit of absorbed dose. It is equivalent to 100 erg g^{-1} (10^{-2} J kg^{-1}). The unit was ratified at the Copenhagen meeting of the International Commission of Radiology in 1953[28].

Gram-roentgen

A unit of absorbed energy[29]. It is equivalent to the energy absorbed (approx. 83·8 erg) when one roentgen is delivered to one gram of air. The unit is very small, absorbed energies being generally of the order of megagram-roentgen (10^3 kg roentgen).

Grashof number (Gr)

This is a non-dimensional parameter used in fluid mechanics which describes the characteristics of free convection. It is equal to $l^3 g \Delta \theta \gamma / v^2$, where l has the dimensions of length, g is the acceleration due to gravity, v is the kinematic viscosity, $\Delta \theta$ is temperature difference and γ is the cubic expansion coefficient. The number occurs in all the laws of free convection. Two convection flows are dynamically similar if their respective Grashof and Prandtl numbers are equal. The number is named after F. Grashof (1826–93) a German authority on heat, whose works included the development of a theory for the draught in a chimney[30]. The number was referred to by name in 1934[31] and was recognized by the American Standards Association[26] in 1941.

The Grashof number (Gr*) used in mass transfer problems is given by

$$l^3 g \left(\frac{\partial \rho}{\partial x} \right) \Delta x \frac{\rho}{\eta^2}$$

where ρ is the density at a point of coordinate x and η is the dynamic viscosity coefficient.

Grave

In 1792 the French government proposed that the mass of a cubic decimetre of water should be called a grave. Seven years later they changed the name to the 'kilogramme' (now generally shortened to kilogram).

Gray (Gy)

The SI unit of absorbed dose of ionizing radiation; 1 gray is equal to an absorption of 1 joule per kilogram, i.e. 1 Gy = 100 rem. The unit which was adopted by the 15th CGPM in 1976, is named after L. H. Gray (1905–65) an eminent authority on cancer and one-time director of the British Empire Cancer Campaign[32]. For radiological protection purposes the absorption of rubber[33] is considered to be about 10^5 Gy, polystyrene 5×10^6 Gy and for ceramics and metals the values lie between 10^8 and 10^9 Gy.

Grey scale
A scale used in colorimetry. (*See* **Surface colour classification.**)

H

Hand
An Anglo-Saxon unit of length. One hand $= \frac{1}{3}$ foot $= 1.016 \times 10^{-1}$ m.

Hardness numbers (solids)
The hardness[1] of solids is described by Mohs, Brinell, Knoop, Meyer, Rockwell, Shore scleroscope and Vickers hardness numbers.

In 1824 F. Mohs[2] devised a scale in which ten minerals are arranged in order of hardness so that each mineral can scratch the ones that come before it on the scale. The original scale consisted of talc (1), gypsum (2), calcite (3), fluorite (4), apatite (5), feldspar (6), quartz (7), topaz (8), corundum (9) and diamond (10). In 1933 Ridgway[3] extended the scale to cover fifteen materials by inserting vitreous pure silica between feldspar and quartz, garnet between quartz and topaz and by replacing corundum by four materials of increasing hardness – fused zirconia (11), fused alumina (12), silicon carbide (13), and boron carbide (14). Diamond (15) remains as the hardest material.

Hardness is also measured by pressing a hard object into the material under test. In this form of test either the area or the depth of the indentation when the object is pressed for a specified time with a specified force indicates the hardness of the material. The earliest form of indentation test was devised by Brinell[4] in 1900. In his test the indenter is a ball of specified diameter – usually 10 mm – and the Brinell hardness number is the ratio of the force applied on the ball to the curved area of the indentation. It is generally expressed in kilogram weight per square millimetre. The Meyer hardness number which dates from 1908 is derived in a somewhat similar manner. Other indentation tests, such as those of Vickers (1922), Rockwell (1922) and Knoop (1939), use steel or diamond cones (pyramids) instead of a ball. In the Rockwell and Knoop tests the depth of the penetration is measured. The Vickers' hardness number originally called Ludwik-Vickers' is sometimes abbreviated to VPN – Vickers pyramid number. In Rockwell tests penetrometers of different sizes depending on the hardness of the

59

material are used and they are designated by letters which follow the appropriate Rockwell number.

In the Shore scleroscope a steel indenter or ball is dropped from a height of 25 cm on to a flat surface of the material concerned. The hardness of the material is given by the height to which the indenter rebounds. The latter is expressed on a scale of 140 equal divisions which starts at zero on the face of the plate and ends at the point from which the indenter was released.

All hardness scales are arbitrary and there is no set of conversions by which the hardness on one scale can be connected to that on the other, but there are countless tables available which give the appropriate hardness numbers for any particular hardness on a specified scale. Thus chilled copper has a Brinell hardness number of 450, a Vickers hardness of 472, a Rockwell number of 46 C, a Shore scleroscope number of 63 and lies between 6 and 7 on Mohs' scale. The Meyer hardness number depends upon the size of the indenter; its value is greater than the Brinell number by a factor which lies between 1 and 2.

Hardness (water)

Hardness of water is now generally expressed as so many parts of calcium carbonate per million parts of liquid, although at one time hardness was expressed in degrees[5]. A French degree represented one part of calcium carbonate in 100 000 parts of water and an English or Clark degree corresponded to one part in 70 000 (i.e. one grain per gallon). On the other hand a German degree represented one part of calcium oxide in 100 000 parts of water so that one German degree was equivalent to 17·8 parts of calcium carbonate per million parts of water.

In England water is said to be soft if its hardness is less than 5° Clark (70 ppm) and very hard when its hardness is greater than 15° Clark (210 ppm). The United States Geological Survey consider water of hardness 0 to 55 ppm to be soft, 55 to 100 to be slightly hard, 101 to 200 to be moderately hard and that above 200 ppm to be very hard. The Clark degree is named after T. Clark (1801–67) who about 1840 devised a satisfactory hardness test for water. The unit dates from the middle of the nineteenth century.

Hartley

A unit of information used with digital computers which is equal to $\log_2 10$ or 3·219 bit.

Hartmann number (Ha)

When a conducting fluid flows in a transverse magnetic field the magnetic forces present oppose viscous action. The Hartmann number[6] is a measure of the relative forces. It is given by $Bl(\kappa/\eta)^{1/2}$ where B is the magnetic flux density, l symbolizes length, κ and η are the coefficients of electrical conductivity and viscosity respectively.

Hartree

A name proposed for a unit of energy in the atomic system of units. It is equal to $4\pi^2 me^4/h^2(4\pi\varepsilon_0)^2$ or (4.3598×10^{-18}) J where m is the rest mass of the electron, e the electron charge, h Planck constant and ε_0 the rationalized permittivity of free space. The unit was named the hartree in 1959[7] after D. R. Hartree (1897–1958) who proposed the atomic units in 1928 (q.v.). Sometimes the value $(4\pi^2 me^4)/h^2$ or 110.5×10^{-21} J is referred to as the Hartree.

Haze factor

A term sometimes used in meteorology. It is defined as B/B', where B' is the luminance of the object and B the luminance of the veil (e.g. mist or fog) through which the object is viewed[8].

Head of liquid

Before 1955 practically all laboratory and meteorological pressure measurements were given in terms of the height of a mercury or water column. Standard, or normal, atmospheric pressure on the mercury scale is represented by a column of mercury of density 13.5951×10^3 kg m^{-3} at 0°C and 760 mm in height when the acceleration due to gravity is 9·80665 m s^{-2}. For low pressures of the order of 10^{-6} mm Hg, the height of the column is frequently given in micrometres. Engineers often express small pressures in terms of a column of water so many inches (or feet) high and on this scale a standard atmosphere would be about 32 feet. Since 1955 meteorologists and physicists have described pressure as force per unit area. (*See* **Standard atmosphere.**)

Heat stress index

See **Wet bulb globe thermometer index.**

Heat transfer factor (j_H)

The quantity St $\times$ Pr$^{2/3}$ is called the heat transfer factor, where St and Pr are the Stanton and Prandtl numbers respectively (q.v.). The factor is used in forced convection[9].

Hectare

A metric unit of area equal to 100 are or 10^4 square metres. It was adopted by the CIPM in 1879. One hectare = 2·47113 acres.

Hedstrom number (He)

This is a non-dimensional parameter used in non-Newtonian fluids. It is given by He $= TD^2\rho/\mu^2$ where T is the yield stress, D the diffusion coefficient, ρ the density and μ the Bingham plastic viscosity of the fluid. It was introduced in 1952[10].

Hefnerkerze (HK)
An absolute unit of luminous intensity used in Germany[11] from 1893 to 1940. 1 hefnerkerze $\simeq 0.9$ international candles.

Hehner number
This number gives the percentage of water and fatty acid in one gram of fat.

Helmholtz
A unit proposed by Guggenheim in 1940[12] for the moment of an electrical double layer. The moment is given by the product of the superficial charge density on each plane and the distance between the planes. The unit is defined as being equal to 1 debye per square ångström which is equivalent to e.s.u. charge per metre. It is named after H. L. F. Helmholtz (1821–94), one of the most versatile of nineteenth century scientists whose interests ranged from physiology to physics and mathematics.

Henry (H)
The henry is the practical and the SI unit of inductance. It may be defined by any one of the well-known relationships $e = -L\mathrm{d}i/\mathrm{d}t$; $\Phi = Li$ or $W = \tfrac{1}{2}Li^2$, where e is the e.m.f. (volts) induced in a circuit, L is the self inductance, $\mathrm{d}i/\mathrm{d}t$ (A sec^{-1}) is the rate of change of the current, Φ (weber) is the total magnetic flux associated with the circuit and W (joule) is the work done in establishing a current i in the circuit. The same definitions hold if M, the coefficient of mutual inductance be substituted for L. In 1948[13] the international henry, which was defined in terms of the international volt and ampere, was replaced by the absolute henry, a unit derived from the absolute volt and the absolute ampere.

The name henry was approved for the unit at the Chigago meeting of the IEC in 1893[14]. It commemorates the name of the American scientist Joseph Henry (1797–1878) who did much of the early work on electrical inductance. Four years previously the IEC adopted the name 'quadrant' for the unit, since the practical unit of inductance is equivalent to 10^9 cm, the length of the earth's quadrant when measured in CGS units. The unit seems to have been called the henry for several years before it received international recognition[15].

(1 international henry $= 1.00049$ absolute henry.)

Herschel
Moon[16] suggested in 1942 that the term radiant power of a source be replaced by the radiant helios of a source and that the unit be called the herschel. It is equal to π times the watts per square metre per steradian and is named after Sir William Herschel (1738–1822), the astronomer who discovered the planet Uranus.

Hertz (Hz)

The SI unit of frequency equal to one cycle per second. It is named after H. R. Hertz (1857–94), the German physicist noted for his experimental work which confirmed J. Clerk Maxwell's electromagnetic theory thereby paving the way for radio. The unit was adopted by the EMMU Committee of the International Electrical Commission[17] in October 1933.

Horse power (hp)

The foot pound unit of power or rate of working. It is equal to 33 000 foot pounds per minute or 745·700 watts. James Watt (1736–1819) invented the unit in 1782 as a selling aid for his steam pumping engines. He assumed the average horse exerted a pull of 180 pounds. Such a horse, when harnessed to a capstan, would walk round a circle of 24 feet diameter about two and a half times a minute. This gave a rate of working of 32 400 foot pounds per minute which Watt rounded off to 33 000 foot pounds per minute. The formal definition of a horse power was published in 1809[18]. In 1889 the British Association[19] expressed the wish that power should be expressed in terms of watts instead of horse power, but little notice was taken of this request for several decades.

The performance of an engine may be described in terms of brake horse power, indicated horse power and nominal horse power. The brake horse power is the power developed which is available for external work. The indicated and nominal horse power are figures derived from the dimensions of the engine. In the case of a piston engine indicated horse power is given by $ASPN/33\,000$ and nominal horse power by $D^2N(S)^{1/3}/15\cdot6$, where A is the area of the piston in square inches, S is the stroke in feet, N is the number of cylinders, P is the pressure on the piston in pounds per square inch and D is the cylinder diameter in inches. From 1921 to 1947 the United Kingdom taxed private motor cars on their 'horse power', this horse power being derived from the Royal Automobile Club formula of $ND^2/2\cdot5$, which however since about 1925 bore no relationship to the power developed by the engine. In the USA the horse power of an automobile is generally given as the brake horse power developed when the engine is running at 4000 revolutions per minute.

Metric horse power (cheval vapeur; pfendstärke) is based on the power required to raise a weight of 75 kg one metre in one second: one metric horse power $= 0\cdot986$ horse power $= 735\cdot5$ watts.

Hour (h)

A unit of time equal to 3600 seconds. In classical times darkness and daylight were each divided into twelve periods the length of which depended on whether it were summer or winter, day or night. The advent of mechanical clocks in the middle ages enabled these periods to represent the same time irrespective of the season of the year, thereby giving 24 hours to the solar day. At one time practically every town in the United Kingdom had its own local time but the passing of the Time Act in 1880[20] imposed Greenwich mean time as the legal standard throughout England and Dublin time (25 minutes behind) in Ireland. The present division of the world into 24 time zones, each about 15° of longitude in width commencing with Greenwich as zero, was proposed by S. Fleming in 1878 and agreed in principle at an international conference held in Washington in 1884[21]. It is not yet adopted everywhere in the world, but is universal throughout Europe, the British Commonwealth and the Americas. Holland was the last European country to adopt the system and, until the 1939–45 war, Dutch time was 20 minutes ahead of Greenwich.

Hubble
This is a unit of astronomical distance equal to 10^9 light years. It was proposed in 1968 and is named after the American astrophysicist E. P. Hubble (1889–1953)[22] who discovered that the majority of extra-galactic nuclei were receding from the earth's own galaxy with velocities proportional to their distances away.

Hundredweight (cwt)
An avoirdupois unit of mass equal to 112 pounds. The 112 pound hundredweight is sometimes called the long hundredweight to distinguish it from the 100 pound (or short) hundredweight which is frequently used in the USA. The hundredweight has been in use in Great Britain since the Tudor times but its value has varied between 100 and 112 pounds.

Hydrogen scale
There is no reliable method for determining the absolute potential of a single electrode, hence electrode potentials are measured with reference to the potential of a reversible hydrogen electrode. This is defined as an electrode with hydrogen as a gas at standard atmospheric pressure in a solution of hydrogen ions at unit activity. The scale was devised by Nernst in 1900[23]. In the original definition the solution was specified as containing one gram ion of hydrogen per litre but the present specification of the solution was introduced in 1913[24]. Zinc has a potential of 0·76 volt when referred to hydrogen, copper (cupric) has a potential of −0·34, hence a cell consisting of a copper and a zinc electrode in an electrolyte such as sulphuric acid develops an e.m.f. of 1·1 volts.

Hydron
A unit of acidity. (*See* **pH index.**)

I

Inch (in)

A unit of length in the foot pound system equal to 1/36 yard. It was agreed by most English-speaking countries in 1959[1] that for scientific purposes the inch would be considered to be equal to 2·54 cm precisely. The inch as a unit of measure can be traced back to medieval times and is derived from the Anglo-Saxon ynce – twelfth part.

Inferno

A unit of stellar temperature suggested in 1968[2]. One inferno is equal to 10^9 kelvin.

Inhour

An arbitrary unit of reactivity proposed in 1947 and used mainly in the USA[3]; if R be the activity of a nuclear reactor and T its period in hours, then R is defined as $1/T$, where R is expressed as inhours.

Instant

A proposed unit of time. (*See* **Degré**.)

International biological standards

The potency of therapeutic substances was at one time expressed in terms of their physiological effects on living animals or organisms. Thus the potency of digitalis was expressed in frog units, which was the amount required to kill one gram of frog. Other animal units in use were cat, dog, guinea-pig, pigeon, mouse, rabbit and rat units. Units such as these were not reliable and have been replaced by a series of units which give for each substance the amount of activity present in an internationally agreed mass (expressed in mg) of the substance concerned, prepared and stored under specified conditions. The substances are known as international biological standards and they are essentially similar to the standard metre and standard kilogram. The units occasionally have names, such as the Voegtlin unit for pituitary extract, but in general they are given as a number representing the mass of the standard. This number is called the International Unit (IU). In the United States the US Pharmacopeia unit (USP) is sometimes used.

The first standard (1922) was the diphtheria antitoxin and the unit is the specific antitoxic activity contained in 0·0628 mg of the anti-toxin and this is expressed by saying one International Unit is equal to 0·0628 mg of the 1st International Standard. Other examples are for insulin for which 1 IU is 0·04167 mg of the 4th International Standard and the penicillin unit (1944) which is the activity contained in 0·0006 mg of the sodium salt of penicillin. A million units of penicillin is a dose of 0·60 g of penicillin salt.

Several score of International Biological standards now exist [4], and these are stored under the specified conditions at the Statens Seruminstitut, Copenhagen, the Central Veterinary Laboratory, Weybridge, England and the National Institute for Medical Research, Mill Hill, London.

Other named units are the Junkmann-Schöller unit for thyrotropin which is 0·1 IU (13·5 mg of the 1st International Standard), the Goldblatt unit (renin and angiotensin), the capon unit (CU) for androgenic activity, Dam and Ansbacher units (vitamin K), the Lipmann unit (pantothenic acid), the avidine unit (biotin) and Lactobacillus lactis dorner unit (vitamin B_{12}). Named units are now seldom used and in most cases nowadays, for example, penicillin, dosages are given in milligrams and not International units. Details are given in Geigy's Scientific Tables[5] and the British National Formulary[6].

International practical temperature scale
See **Temperature scales.**

Intrinsic viscosity
See **Limiting viscosity number.**

Ionic strength (μ)
This is a measure of the intensity of the electrical field existing in a solution. Ionic strength is defined as half the sum of the products of ion molalities and the square of the ion valencies. For a univalent electrolyte μ is the molality; for a bivalent substance it is four times the molality and for an electrolyte with one bivalent ion and one univalent ion μ is three times the molality. The term was first introduced by Lewis and Randall in 1921[7].

Jansky

A unit used in radio astronomy to indicate the flux density of electromagnetic radiation received from outer space, [1 and 2] one jansky being equal to 10^{-26} W m^{-2} Hz^{-1}. The unit is named after the American electrical engineer Karl G. Jansky (1905–1950), who became the first radio astronomer in the world in December 1930 when he detected electromagnetic radiation of wavelength 15 m coming from the Milky Way (*see* **Flux unit**). It was adopted in August 1973 by the International Astronomical Union.

Jar

A unit of capacitance equal to 10^3 cm stat units. 9×10^8 jars = 1 farad. The unit represented the approximate capacitance of an early Leiden jar and was used quite extensively by early experimenters in the middle of the nineteenth century. It was used at one time by the Royal Navy and is defined in all the editions of the *Admiralty Handbook of Wireless Telegraphy*[3] published between 1920 and 1938. The jar is probably the oldest electrical unit, coming into use about 1834 when it was introduced by Sir William S. Harris[4] (1792–1867). However, the idea of using the charge stored in a jar as a quantitative unit had been adopted many years before this and is reputed to have been devised by Benjamin Franklin in the middle of the eighteenth century. The unit is now obsolete.

Jerk

A unit at one time used by some engineers to express the rate of change in acceleration, where 1 jerk was equal to an acceleration change of 1 foot per second^2.

Joule (J)

The SI and the practical unit of work which is equal to a force of one newton acting over a distance of one metre. It is equivalent to 10^7 ergs and in electrical units is the energy dissipated by 1 watt in a second. The unit was originally proposed by the British Association in 1888[5] who recommended that it be named after J. P. Joule (1818–89), the originator of the mechanical theory of heat. It was recognized by the IEC in 1889. In 1948 the joule[6] was adopted as the unit of heat by the International Conference on Weights and Measures, so that the specific heat of water at 15°C is 4185·5 joule (kg °C)$^{-1}$, a figure previously always associated with J, the mechanical equivalent of heat. The idea of using the joule as a unit of heat had the support of the British Association[7] in 1896, but little notice was taken of their proposal.

$$(1 \text{ J} = 2 \cdot 78 \times 10^{-7} \text{ kWh} = 9 \cdot 47 \times 10^{-4} \text{ Btu} = 9 \cdot 47 \times 10^{-9} \text{ therm.})$$

K

K factor
The name given to the γ ray dose rate in roentgens per hour at a distance of 1 centimetre from a 1 millicurie point source of radiation[1]. Each γ emitter has its own K factor.

Kanne
An alternative name[2] for the litre (q.v.).

Kapp line
A line of magnetic induction in which each line represents a flux of 6000 maxwells. The unit was devised by Gisbert Kapp (1852–1922) in 1886 who named the line after himself[3].

Kayser
In spectroscopy the frequency of electromagnetic radiation is generally expressed as a wave number, which is the reciprocal of wavelength of the radiation concerned. if the wavelength be expressed in centimetres, the reciprocal gives the number of wavelengths in a centimetre. This figure has the dimensions of cm^{-1} and is expressed in kaysers. The name was approved for the unit of wave number in 1952[4] to commemorate J. H. G. Kayser (1853–1940) the compiler of the great spectra catalogue *Handbuch der Spectroscopie*. Other names which have been suggested for the unit are the balmer[5] and the rydberg[6] but neither has been adopted.

The convenience of using reciprocal wavelengths in spectroscopic calculations was originally pointed out by George J. Stoney in 1871[7] but little notice was taken until the idea was used by Hartley[8] in 1883. The energy represented by 1 kayser is $123 \cdot 9766 \times 10^{-6}$ electron volt.

Kelvin
A name occasionally used for the kilowatt hour (q.v.).

Kelvin (K)

The kelvin is the fundamental unit of temperature in the SI system. It is defined as the fraction 1/273·16 of the thermodynamic temperature of the triple point of water. The definition was adopted by the 16th CGPM (1968) when it was decided to change the name of the unit from the degree kelvin (°K) to the kelvin (K)[9]. It is named after Lord Kelvin (1824–1907), one of the outstanding physicists of the nineteenth century.

Kerma

The name used in radiology to describe the kinetic energy transferred to charged particles in unit mass of material by uncharged particles, e.g. neutrons[10]. The kerma may be expressed in either $J\ kg^{-1}$ or $erg\ g^{-1}$; the name is derived from the initial letters of *k*inetic *e*nergy *r*eleased in *ma*terial; it was authorized by the ICRU in 1962.

Kilocalorie

Physiologists when discussing metabolism express their findings in calories but they actually mean kilocalories. Thus the statement that eating two fish balls is equivalent to absorbing 250 calories means that during an experiment carried out in a bomb calorimeter, 250 kilocalories of heat are released when the balls are burnt up in an atmosphere of pure oxygen.

Kilogram (kg)

The fundamental unit of mass in the metric system. It was originally supposed to represent the mass of a cubic decimetre of water at the temperature of its maximum density. Subsequent work showed that the cube used for the determination of the kilogram of 1799, known as the Kilogramme des Archives, had a volume of 1·000028 cubic decimetres, thereby making it too heavy. At the Convention du Metre held in Paris[11] in May 1875 it was decided to make a new international kilogram which would be an exact copy of the Kilogramme des Archives. The new standard, known as the International Prototype Kilogramme was established in 1889[12] and copies of it were sent to the various bodies responsible for weights and measures throughout the world. Copy No. 18 was sent to England, No. 20 to the USA.

In the British Parliament during the 1962–63 Session[13] a weights and measures Bill was considered in which the Imperial pound was defined in the following terms: 'the pound shall be 0·45359237 kilograms exactly'. This Act, passed in 1963, makes the kilogram the fundamental unit of mass in both the metric and the foot pound second systems of units.

Kilogram equivalent

The kilogram equivalent mass of an element or radical is equal to its kilogram atomic weight divided by its valency.

Kilowatt hour (kWh)

The kilowatt hour is the commercial unit by which electricity is sold to the consumer. The unit was first specified in the Board of Trade Orders made under the 1882 Electricity Act[14]. Its original definition is given as the 'energy contained in a current of one thousand amperes flowing under an electromotive force of one volt during one hour'. Other names for the unit are the Board of Trade unit and the kelvin.

Kine

The name suggested by the British Association[5] in 1888 for the CGS unit of velocity. It has never been widely used but Firestone[16] used it in 1933 in a definition of the mechanical ohm.

Kip

A unit of mass used mainly by engineers for expressing the load on a structure. A kip is equal to 1000 pounds and probably derives its name from the initial letters of '*K*ilo *I*mperial *P*ounds'. It came into use in the USA before 1939. Its first appearance in a book published in England was in 1948[17].

Knot (kn)

A speed equal to 1 nautical mile (6080 ft) per hour[18]. The unit dates from the late sixteenth century, when the speed of a ship was found by dropping a float tied to a knotted line (knotted log) over the side of the vessel. The knots were originally 7 fathoms apart and the number of knots passing in 30 seconds gave the speed of the ship in 'nautical miles' per hour. Subsequent surveys altered the length of the nautical mile and by the beginning of the nineteenth century the knots were about four fathoms apart and the number of these passing in 14 seconds gave the speed. Some sea captains, however, preferred the knots to be 8 fathoms apart, in which case the standard time was 28 seconds. Special 14 and 28 second hour glasses were available for timing the log.

There are actually two knots in use, the international knot of 1852 metre hour^{-1} and the United Kingdom knot of 6080 feet hour^{-1}. 1 UK knot = 1·00064 international knots. The USA adopted the international knot in July 1954.

Knoop scale

A hardness scale. (*See* **Hardness numbers**.)

Knudsen number (Kn)

The Knudsen number[19] is connected with the flow of gases at very low pressure when the mean free paths of the molecules are of the same order of magnitude as the dimensions of the path along which they are flowing. The Knudsen number is given by λ/l, where λ is the mean free path of the molecules, l is a length derived from the dimensions of the apparatus. The number is named after M. H. C. Knudsen (1871–1949) the inventor of the gauge for the determination of low pressures which carries his name.

Konig

A name suggested in 1946 for the X stimulus in the trichromatic colour system. (*See* **Trichromatic unit.**)

Krebs unit

A unit used as a measure of consistency, particularly with reference to paints[20].

Kunitz unit

A unit used in biochemistry to describe the concentration or activity of the enzyme ribonuclease. One kunitz is the amount of ribonuclease required to cause a decrease of 100 per cent per minute in the ultraviolet light (300 nm) absorbed at 25°C by a 0·05 per cent solution of yeast nucleic acid in a 0·05 molar acetate buffer solution (pH 5·0). It is named after the Russian-born American biochemist M. Kunitz (1887–   ) who proposed the unit in 1946.

L

Lambda

A unit of volume used mainly in microchemistry. 1 lambda is equal to a microlitre. It was proposed in 1933[1]. The name of the Greek equivalent of the letter L seems to have been used for the unit as a shorthand form of μl – microlitre.

Lambert (L)

A unit of luminance of a surface equal to 1 lumen per square centimetre. The unit is rather large for normal photometric work and the millilambert is frequently used instead. The millilambert (0·001 lambert) was at one time employed extensively in the USA, but in England the foot candle was preferred. (1 foot candle is equivalent to 1·0764 millilamberts.) Luminance expressed in candles cm^{-2} or candles in^{-2} may be reduced to lamberts by multiplying by π or $\pi/6·45$ respectively.

Since 1949[2] the lambert has been replaced by the apostilb which is 1 lumen per square metre. Neither the lambert nor the apostilb are recommended for scientific work where it is preferable to express luminance in terms of candela per unit area. The lambert is named after J. H. Lambert (1728–77) an early worker on photometry; the unit was first used about 1920.

Langley

The langley was put forward as the unit of solar radiation density in 1942; it is defined in terms of the 15°C gram calorie per square centimetre per minute, or 697·8 W m^{-2} s^{-1}. As the mean total solar radiation is about 392×10^{24} W, the average energy received at the surface of the earth is about $1·3 \times 10^3$ W m^{-2} or 2 langleys. A proposal was made in 1947[3] to change the dimensions to W m^{-2}, thereby leaving the time to be specified; thus the mean solar radiation density at the earth's surface would be expressed as 2 langley $minute^{-1}$. The unit was proposed by Linke[4] and is named after S. P. Langley (1834–1906) the first director of the Astrophysical Laboratory at the Smithsonian Institution. (*See* **Pyron**.)

League
This was widely used throughout Europe as an itinerant unit of distance. Its value, which varied slightly from place to place, was about three miles.

Lentor
A name used between 1920 and 1940 for the CGS unit of kinematic viscosity. It is now called the stokes (q.v.).

Leo
A metric unit of acceleration representing an acceleration of 1 decametre per second per second. The unit was seldom used, but it has a place in Glazebrook's *Dictionary of Applied Physics* (1922)[5].

Lewis number (Le)
The Lewis number is the ratio of the diffusivity to the diffusion coefficient of a fluid, where the diffusivity K is defined as $k/c\rho$ in which k is thermal conductivity, c is specific heat and ρ density, and the diffusion coefficient has the dimensions of length^2 time^{-1}. For gases the Lewis number varies from 0·8 to 1·2 and for liquids it lies within the range 70 to 100. The name for the number was proposed[6] in 1948 but the idea of the number was mooted by G. W. Lewis[7] in 1939.

Light watt
The luminous efficiency L of a source of radiation is the number of lumens associated with 1 watt of radiant power. However, this efficiency has to be evaluated in terms of the brightness sensation it produces to the eye and since the latter varies with the wavelength, radiant power is expressed in light watts. A light watt is equivalent to $1/V_\lambda$ watts, where V_λ is the reciprocal of the relative amount of power required to produce a given brightness sensation. The values of V_λ, which is known as the relative luminous efficiency of light (formerly relative visibility) are given for the light-adapted or photopic eye in standard tables[8]. The maximum value occurs at 555 nm. For the dark-adapted or scotopic eye the values are slightly different with the maximum at about 510 nm. The maximum value of V_λ corresponds to 680 lumens associated with 1 watt and the reciprocal of this maximum absolute value, i.e. 0·00147 watts per lumen at 555 nm is called the mechanical equivalent of light. The relationship between L and V_λ for a source having a continuous energy distribution represented by E_λ is

$$L = 680 \int_0^\infty V_\lambda E_\lambda \mathrm{d}\lambda \bigg/ \int_0^\infty E_\lambda \mathrm{d}\lambda,$$

which for a monochromatic source is reduced to $L = 680\, V_\lambda$. In the latter case a source of luminous efficiency L lumens per watt has a radiant power of $680/L$ light watts.

Light year (ly)

A unit used in popular astronomical literature to describe stellar distances. 1 light year is the distance traversed by electromagnetic radiation in a year; this is equal to $9·4605 \times 10^{15}$ metres or about $0·33$ parsec. The unit was used for the first time[9] in 1888.

Limiting viscosity number (LVN)

A number characterizing the viscosity of a suspension. It is defined as the limiting value of η_{sp}/c as the concentration c approaches zero. η_{sp} is the specific viscosity, i.e. the value of $(\eta_s - \eta_0)/\eta_0$ in which η_s and η_0 are the dynamic viscosity coefficients of the suspension and suspending medium or solvent respectively. The name was recommended as a replacement for the term intrinsic viscosity by the Commission on Macro-molecules of the International Union of Pure and Applied Chemistry in 1951[10].

Line

The line is a unit of length equal to 1/12 inch. It was in use by the seventeenth century and today is often employed by botanists in describing the size of plants. Its most recent appearance in a standard work is in the 1954 reprint of the 7th Edition of Betham and Hookers' *Handbook of the British Flora*. The Paris line (or Ligne) is 1/12 of a Paris inch. The line used in the USA is 1/40 inch.

Line of induction

A line of magnetic induction is a conception for describing magnetic flux introduced by Faraday (1791–1867). 1 line is equivalent to 1 CGS unit of flux, i.e. 1 maxwell. A magnetic flux of 10^6 maxwells is 1 megaline and a flux of 6000 maxwells is a kappline (q.v.).

Lines of electrostatic induction are used for describing the intensity of electric flux: 1 line is equivalent to a flux of 1 CGS electrostatic unit of charge.

Link

An Anglo-Saxon unit of length.

$$\text{One link} = 0·66 \text{ foot} = 2·01168 \times 10^{-1} \text{ m.}$$

Litre (l or L)

The litre is the subsidiary unit of volume in the metric system. It is defined as the volume occupied by 1 kilogram of water at its temperature of maximum density; it is equal to $1·000028$ cubic decimetres (approximately $0·22$ gallon).

The litre dates from the French Revolution. It is unique in that it is the only metric unit to carry a pre-revolutionary name (the litron was the French Royalist unit of volume). The litre was chosen to represent the volume occupied by a kilogram of water and was originally supposed to be the same volume as a cubic decimetre. Subsequent work showed the kilogram from which it was derived was too large. The discrepancy between the litre and the cubic decimetre was recognized at the Conférence du Metre in 1875 but no international action was taken until the 1901 meeting of the International Weights and Measures Congress[11] when the volume of a litre was confirmed as being 1·000028 cubic decimetres. In 1964 the CIPM repealed this definition and abolished the litre as a precisely defined unit of volume[12]. It suggested that for approximate measurement of volume the litre could be used as the name for the cubic decimetre; the inaccuracy involved would be of the order of 1 in 36 000.

When the unit was first introduced in 1793 it was called the pinte but this name was replaced by 'litre' in 1795.

The litre was recognized as a unit of volume in England by an Order in Council dated 19 May 1890.

Logarithmic scales of pressure

Townsend in 1945 suggested[13] a logarithmic scale for describing low pressure. Since a given pressure can be expressed as a fraction f of a reference pressure, say 1 millimetre of mercury, then pressure can be indicated by a figure obtained by multiplying by minus ten the mantissa of the logarithm to base ten of f. Thus 10^{-2} mm Hg gives a mantissa of -2 corresponding to a pressure of 20 units on this scale. No name is required for the unit as the number itself is a sufficiently clear indication of the pressure. Pressures greater than 1 mm Hg would have a negative value on the Townsend logarithmic scale. Rose[14] proposed a similar scale, but he called his unit a decilog thereby making the decilog in pressure analogous to the decibel in sound. Denne[15] has also supported a logarithmic pressure scale.

In the Boyle scale, pressure is expressed with respect to a unit pressure of 1 bar (10^5 N m^{-2}). Thus the relationship

$$dB = 10 \log_{10} p - 28·75$$

expresses a pressure of p torr in terms of the deciboyle (dB); if p be in atmospheres the same formula holds provided $+0·057$ is substituted for $-28·75$. In this scale 10^{-11} torr is expressed as -140 dB and the critical pressure of air (33·7 atmospheres) as $+15·7$ dB. The scale was proposed in 1964 and is named after the Hon. Robert Boyle (1627–91) of Boyle's law fame[16].

Logit

A name proposed in 1952 for the decibel (q.v.).

Lorentz unit

A spectral line of frequency f splits into components having different frequencies when the light source is placed in a magnetic field. In the normal Zeeman effect the lines are split into two components of frequencies $f \pm \Delta f$ when seen in a direction along the field and three components of frequencies

$f, f \pm \Delta f$ when seen transversely. The frequency difference Δf in wave numbers, i.e. $\Delta f/c$, is known as the Lorentz unit. The value of the unit is $\mu_0 e/4\pi mc$ metre^{-1} per unit field in MKS units and $e/4\pi mc^2$ cm^{-1} or $4 \cdot 67 \times 10^{-5}$ cm^{-1} in e.m.u., where μ_0 is the permeability of free space, e the electronic charge, c the velocity of light and m the electron rest mass. The Lorentz unit is equal to the Bohr magneton expressed in wave numbers. It is named after H. A. Lorentz (1853–1928), professor of mathematical physics at Leiden. The unit is seldom used but was employed by White in his book on atomic spectra[17] in 1934.

Loschmidt number or Loschmidt constant (N_L)
This is the number of molecules in a cubic metre of an ideal gas at s.t.p. It is equal to the Avogadro contant divided by the molar volume and its value is $2 \cdot 6868 \times 10^{25}$ m^{-3}. The number was first derived by J. Loschmidt (1821–95) in Vienna in 1865. In Germany it is sometimes called Avogadro's constant.

Loudness unit (LU)
A unit proposed in 1937[18] and adopted by the American Standards Association in 1942[19]. One loudness unit was defined as corresponding to a loudness level of zero phon. As there was no mention of frequency in the definition the object of the unit, which was to give a loudness scale based on the loudness heard by the average listener, was not achieved. The loudness unit has been replaced by the sone, where 1 millisone corresponds to 1 loudness unit.

Lumberg
A CGS unit of luminous intensity. An erg of radiant energy which has a luminous efficiency of x lumens per watt has its luminous energy described as x lumbergs. The term came into use just before the 1939–45 war. An earlier name suggested by the American Optical Society in 1937 was the lumerg[20].

Lumen (lm)
The lumen is the SI unit by which the rate of flow (flux) of luminous energy is evaluated in terms of its visual effect. It is the luminous flux emitted from a point source of uniform intensity of 1 candela into unit solid angle so that the total flux from 1 candela is 12·57 lumens. 1 lumen at 5550 Å equals 0·001 470 5882 W. It has been a legal unit in France since 1919[21], but it is claimed that A. Blondel (1863–1938) proposed the unit as long ago as 1894[22] when the source was 1 international candle.

Lumen-hour
A unit of quantity of light. It is equal to a flux of 1 lumen continued for 1 hour. This unit was adopted about 1920[23] by the Illuminating

Engineering Society of New York and by the American Engineering Standards Committee. Sometimes the lumen-second is used.

Lumerg
An early name for the lumberg (q.v.).

Lundquist number
A number used in magnetohydrodynamics[24] to characterize unidirectional Alfvén waves, i.e. waves set up in a conducting fluid flowing in a magnetic field. It is given by $B\sigma l(\mu/\rho)^{1/2}$ where B is the magnetic flux density, σ is electrical conductivity, μ is permeability, ρ is density and l symbolizes length.

Lunge
A specific gravity unit. (*See* **Degree (hydrometry)**.)

Lusec
A unit giving the speed of pumping of a vacuum pump. It is equal to 1 litre per second at a pressure of 1 micrometre. The name originated after 1945.

Lux (lx)
The SI Unit of illumination which is equal to 1 lumen per square metre. The unit was introduced in Germany[25] in 1897 where it was associated with the metre-hefner and at one time it was called the metre-candle. The corresponding British unit of illumination is the lumen per square foot or the foot-candle and this is equivalent to 10·764 lux. The lux-sec is sometimes used as an MKS unit of exposure[22].

Luxon
The Luxon is the retinal illumination produced by a surface having a luminance of 1 candela per square metre when the area of the aperture of the eye is 1 square millimetre. The unit was first proposed in 1916 by L. T. Troland who called it a photon[26]. It is also called a troland (q.v.).

M

Mach number (Ma)
The mach number is used frequently in aerodynamics. It is the ratio of the speed of an object compared with the speed of sound in the same medium. The number owes its origin to the Austrian scientist E. Mach (1838–1916) who used it in 1887[1]. It is also employed in ballistics and in heat-transfer work.

Mache unit
The quantity of radioactive emanation which sets up a saturation current equal to 10^{-3} stat units of current. 1 mache unit is equal to 3.6×10^{-10} curie. The unit was defined by the International Radium Standards Committee in 1930[2]. It is named after H. Mache (1876–1954) and is now obsolete.

MacMichael degree
A unit used in viscometry. (*See* **Degree (viscometry).**)

McLeod
A name suggested in 1945 for a unit of pressure on a logarithmic scale[3]. Pressure A in McLeods is given by $A = -\log_{10} p$, where p is the pressure in millimetres of mercury. It is named after H. McLeod (1841–1923) the inventor of the McLeod vacuum gauge.

Madelung constant (a)
A constant used in calculating Coulomb energy and of paramount importance in ionic crystal calculations. In calculating the constant a reference ion is taken; then, if r_j is the distance of the jth ion from the reference ion and R is the nearest-neighbour distance, the Madelung constant is given by

$$\frac{\alpha}{R} = \sum \frac{(\pm \text{charges})}{r_j}$$

Typical values of the Madelung constant based on unit charges and referred to nearest-neighbour distances are 1·7476, 1·7626, 1·6381 for sodium chloride, caesium chloride and zinc sulphide respectively[4].

Magic number

Magic numbers are the names given to numbers which signify the number of electrons in atoms of unusual stability or the number of protons and/or neutrons in very stable nuclei. For atoms the numbers are 2, 10, 18, 36, 54, 86, which are the atomic numbers of the noble gases helium, neon, argon, krypton, xenon and radon. For nuclei the numbers are 2, 8, 20, 28, 50, 82 and 126. The term magic number was first used in connection with nuclei in 1949[5].

Magnetic Reynolds' number (Re_m)

A number used in magnetohydrodynamics to indicate the relative magnitudes of convection and diffusion of the magnetic field. It is given by $v\mu\kappa l$ where v is velocity, μ is permeability, κ is electrical conductivity and l has the dimensions of length.

Magneton

A unit of magnetic moment used in atomic physics which is called the Bohr magneton or nuclear magneton according to whether it refers to an electron or to a proton. The Bohr magneton is equal to $eh/4\pi m$ ($9\cdot274 \times 10^{-24}$ JT^{-1}) in rationalized MKS units and to $eh/4\pi mc$ ($9\cdot274 \times 10^{-21}$ erg oersted^{-1}) in e.m.u., where e is the charge of the electron, h is Planck's constant, m is the rest mass of the electron, and c the velocity of light. The nuclear magneton is approximately 1/1836 of the Bohr magneton and is equal to $eh/4\pi M$ ($5\cdot051 \times 10^{-27}$ JT^{-1}) in rationalized MKS units and $eh/4\pi Mc$ ($5\cdot051 \times 10^{-24}$ erg oersted^{-1}) in e.m.u., where M is the rest mass of the proton and e, h and c as defined above.

The expression $eh/4\pi mc$ was first used by Bohr in connection with the Zeeman effect in 1914. The name magneton was devised by Weiss[6] in 1911 but was first used with specific reference to $eh/4\pi mc$ by Professor H. S. Allen in 1915[7]. The term Bohr magneton first appeared in print in 1925[8]. The term nuclear magneton was first use in 1935[9] but previous to this date it was sometimes referred to as the nuclear Bohr magneton[10].

Marangoni number

This number is named after the Florentine mathematician C. G. Marangoni (1840–1926); it is used to indicate the interdependence of temperature, surface tension and convection in regulating the flow of heat through thin surface layers of fluid.

Margulis number

A number used in heat transfer calculations; the product of the Peclet number (q.v.) and the Margulis number is equal to the Nusselt number (q.v.).

Mass transfer factor (j_M)

This is a dimensionless constant used in heat transfer.

Maxwell (Mx)

The maxwell is the CGS unit of magnetic flux. It is a small unit, for if the magnetic flux linked with a coil of N turns be changing at a steady rate of 1 maxwell per second, a potential difference of N abvolts ($N \times 10^{-8}$ practical volts) is developed in the coil. 1 maxwell is equivalent to one line of induction – see lines of induction.

The name maxwell (after James Clerk Maxwell (1831–79), the originator of electromagnetic theory) was adopted by the IEC in 1900[11]. The idea of giving internationally recognized names to CGS electrical units had been discussed by the same body eleven years earlier. The name maxwell was confirmed in 1930[12].

Mayer

A unit of heat capacity used in 1925 by Richards and Glucker[13]. A substance has a heat capacity of 1 mayer if 1 joule raises the temperature of 1 gram by 1 degree Celsius, so that water at 10°C has a heat capacity of 4·1902 mayers. The unit is named after J. R. Mayer (1814–78) an early worker on the correlation between heat and energy.

Mechanical ohm

The mechanical ohm is the unit of mechanical impedance. The mechanical impedance of a surface is defined as the ratio of the effective sound pressure on a specified area of an acoustic medium to the resulting linear velocity through it. This ratio is a complex number. The dimensions of mechanical impedance are dyne second cm^{-1} in the CGS system and newton second metre^{-1} in MKS units. Modern practice favours the use of the MKS unit which is designated the MKS mechanical ohm. The mechanical ohm was proposed by Firestone[14] in 1933, and he gave the dimensions of the unit as dyne/kine, this being one of the few instances in which the name kine has been used for the CGS unit of velocity.

Megaton

A unit used to define the magnitude of an explosion equal to the detonation of a million tons of TNT. The megaton came into use about 1950 when it was used to describe the force of explosion of a hydrogen bomb. From 1945 to 1960 atomic explosions were described in terms of kilotons of TNT. One standard atomic bomb was considered to be equivalent to 20 000 tons of TNT; this corresponds to about 10^{15} kilojoule.

Mel

A unit of subjective pitch. A simple tone of frequency 1000 Hz which is 60 dB above the listener's threshold (0·0002 μB) gives a pitch of 1000 mel. The unit was initially defined in 1937[15] and was approved as an American standard in 1951[16]. The name mel is derived from the first three letters of *mel*ody.

Met

The unit of metabolism in which 1 met is equal to the metabolism for a seated resting person. It equals 58·15 W m^{-2}. Met values for a number of activities are as follows: lying down 0·8; sedentary activity 1·2; standing relaxed 1·2; light activity standing 1·6; medium activity standing 2·0; high activity 3·0.

Metre (m)

The metre is the SI base unit of length. It was defined at the 17th CGPM (1984) as the length of the path travelled by light in vacuum during a time interval of 1/(299 792 458) second[17]. This definition does not change the size of the unit but it was introduced to take into account recent developments in measurement techniques whereby length and time can be reproduced with very high accuracy, in the case of the second to an accuracy better than one in 10^{13}. An outcome of the new definition is that the speed of light has become the second physical constant to be fixed by convention, the other being the permeability of free space (*see* **Speed of light**).

The metre had its origin in August 1793 when the Republican Government in France decreed the unit of length would be 10^{-7} of the earth's quadrant passing through Paris and that the unit be called the metre. Five years later the survey of the arc was completed and three platinum standards and several iron copies of the metre were made. Subsequent examination showed that the length of the earth's quadrant had been wrongly surveyed but instead of altering the length of the metre to maintain the 10^{-7} ratio, the metre was redefined as the distance between two marks on a bar.

In the nineteenth century scientists throughout the world gradually adopted the metric system for their measurements. This led to the standardization of the metre which was ratified at the Metric Conference in Paris in 1875. The present international prototype metre, which was established in 1889, was a direct consequence of this conference. During the century many of the non-metric countries passed the necessary legislation to define the metre in terms of their own units of length. Thus the United States Congress defined the metre in terms of the US yard and in 1866 the British Parliament, through the Weights and Measures Act of 1866, permitted the metric system to be used for contracts but not for trade. Scientists, however, were not satisfied at having to use an arbitrary standard as their fundamental unit. As early as 1828 Babinet[18] suggested the wavelength of light as a natural unit of length but over a hundred and thirty years had to elapse before the wavelength unit was adopted. In the intervening years two serious attempts were made to establish an optical unit of length. One was by Michelson in 1895 when he suggested the red line of cadmium; the other took place in 1907 when spectroscopists defined their unit – the ångström – in terms of this line which made the ångström equal to $1.0000002 \times 10^{-10}$ m, whereas it was meant to represent a distance of 10^{-10} m exactly. This discrepancy was overcome when the metre was redefined in terms of the wavelength of the orange line of krypton at the 11th CGPM (1960). In the same year a proposal was before the British Parliament to define the imperial standard yard in terms of the metre so that 1 yard $= 0.9144$ metre exactly. Pressure of parliamentary business delayed the passing of the necessary act[19] until 1963, since which time the metre has been the basic unit of length in both the metric and imperial system of units.

Metre atmosphere
(*See* **Atmo-metre**)

Metric slug
A name suggested[20] in 1939 for the unit of mass in the metric (kilogram weight) system of units. It is the mass which is accelerated by 1 metre second $^{-2}$ by a force of 1 kilogram weight; it is equal to g kilograms (approximately 9.8 kg). It is sometimes called a hyl, a mug, a par, or a TME.

MeV
An abbreviation used to express energy in units of 10^6 electron volts (q.v.).

Meyer number
A hardness number (q.v.).

Mho
The mho is the unit of conductance or reciprocal of impedance. The term is not recognized internationally but it is extensively used to express the admittance of circuits containing inductance, capacitance and resistance. A circuit of impedance x ohms has a conductance of $1/x$ mhos. In recent years attempts have been made to replace the mho by the siemens for circuits containing only non-inductive resistance.

The name mho was first used by Lord Kelvin at a meeting of Civil Engineers on 3 May 1883[21]. In 1922[22] Cumming and Kay used the name gemmho for the reciprocal megohm. The conductance of distilled water is of the order of 3 gemmhos.

Mic

A name suggested for a unit of inductance which is equal to 10^{-6} henry. It appears in all editions of the *Admiralty Handbook of Wireless Telegraphy* between 1920 and 1938[23], but it seems never to have been used outside the Royal Navy.

Michaelis constant (K_m)

In enzyme assays measurements of initial rates of reaction should be used. To facilitate the measurement of initial rates the substrate concentration should be high enough to saturate the enzyme. If this concentration is too low a constant is required to convert the observed rate into that which would be obtained on saturation of the enzyme with substrate. This constant is known as the Michaelis constant, K_m. It is given by $V = v(1 + K_m/c)$, where v is the observed rate at substrate concentration c and V is the rate at saturation.

Micri-erg

A name proposed by Harkins[24] in 1922 for a unit of energy equal to 10^{-14} erg, it was used for describing the surface energy of molecules.

Micril

A name suggested for the gammil (q.v.).

Micron μ (length)

A unit of length equal to 10^{-6} m. It was approved by the CIPM in 1879 and was in general use by the 1890s[25], its abbreviation μ appearing in a Board of Trade report in 1895[26]. In 1968 the 13th CGPM decided to proscribe the use of the term micron and express 10^{-6} m as the micrometre (μm)[27].

Micron (pressure)

A unit of pressure equal to 10^{-6} metre or 10^{-3} mm Hg. It is equal to approximately one millitorr. (*See* **Head of liquid**.)

Mil (angular)

The United States artillery in World War II used a system of angular measure in which a right angle was divided into 1000 parts called mils[28]. 1 mil was approximately equal to the angle subtended by a yard at a range of a thousand yards. A mil has also been used to indicate a thousandth of a radian[29].

Mil (circular)

The area of a circle of 0·001 inch diameter. One circular mil = 10^{-6} circular inch.

Mil (length)

A unit of length equal to 1/1000 inch. The name was given to the unit by James Cocker (Liverpool) in 1858[30] but did not come into general use until after its appearance in *The Journal of the Institution of Telegraph Engineers* in 1872[31]. A thousandth of an inch is also called a thou.

Mil (volume)

An abbreviation for millilitre used mainly in pharmacy. It was authorized by the Board of Trade[32] as an official name for the millilitre in 1905.

Mile (geographical)

The distance subtended by 4 minutes of arc on the equator: it equals 7421·591 m.

Mile (nautical) (n. mile)

The nautical mile is the average meridian length of 1 minute of latitude. The British Admiralty and Mercantile marine until recently used the 6080 ft nautical mile (1 minute of arc at 48° latitude) but now in agreement with all other nations use a nautical mile of 1852 m (6076·115 ft) (1 minute of arc at 45°), the distance recommended by the International Hydrographic Conference in 1929; 1 English nautical mile = 1·00064 international nautical miles. A telegraph nautical mile is 6087 feet[33], which is the length of a minute of arc at the equator.

The nautical mile came into use in the early seventeenth century. The idea of using the distance subtended by a minute of arc was suggested by E. Gunter[34] (1581–1626) to help navigators transform the astronomical distances subtended by heavenly bodies to distances on the surface of the sea. It is also called the sea mile.

Mile (land or statute)

The English mile of 1760 yards was authorized in 1592. Its origin seems to have been the Roman marching unit of 1000 double paces – a double pace was 5 feet. At one time nearly every part of England had its own mile which in some instances was as long as 2880 yards. The 2240 yard mile survived in parts of Ireland until well into the present century.

Miller indices

These are integers which determine the position of a crystal plane with reference to three crystallographic axes which are mutually at right angles. The reciprocals of the intercepts of the plane on the three axes are expressed in terms of the lattice constants and are then reduced to the smallest three integers having the same ratio. These integers are the Miller indices which were proposed by W. H. Miller[35] (1801–80) in 1839. Thus if the intercepts are $(3a, 0, 0)$, $(0, 2b, 0)$ and $(0, 0, 4c)$ where a, b, and c are lattice constants, the reciprocals are 1/3, 1/2 and 1/4 so the miller indices are 4, 6 and 3.

Millicuries – destroyed (mcd)
A unit of X-ray dosage which originated in France about 1920. It represented a dose equivalent to that emitted from a radioactive source during the time its radioactivity falls by a millicurie[36]. It was used mainly with radon for which the mcd value was 133 milligram hour.

Millier
A name used in the 1878 Weights and Measures Act to represent a million grams.

Millihg
A unit of pressure equal to 1 mm Hg, or approximately one torr (q.v.).

Milton Keynes energy cost index
An energy performance standard relating to buildings. Based on an estimate of a house's total annual energy running costs per square metre under certain standard conditions of occupancy and use. The index varies between 90 and 250, the lower the figure the lower the energy running costs. Typical three-bedroomed houses built to British building regulations have an index value of about 170, most mid-European countries have an index between 140 and 170 whereas Scandinavian houses have values between 100 and 110.

Minimal erythema dose (MED)
This unit refers to the reddening of the skin in sunburn and is a unit that can be used either as measured by instruments or as calculated from very detailed questionnaires as a measure of ultra-violet dosage. it is considered that 600 MEDs per year over a period of 30 years will give skin cancer.

Minute
The minute of arc and the minute of time are survivals from the Babylonian system of working in units of sixty. There are 60 minutes in a degree and 60 minutes in an hour. So far attempts to decimalize units of angular measure and of time have been particularly unsuccessful.

Mired
Colour temperature is sometimes expressed in micro-reciprocal degrees (Kelvin); this unit is called the mired. Thus a temperature of 2000 K corresponds to a reciprocal of 500×10^{-6} and is equal to 500 mired while 50 000 K is equivalent to 20 mired. The idea of using micro-reciprocal degrees was suggested by Priest in 1933[37] to give reasonably sized figures

for expressing high tempeatures, the name itself being derived from the first two letters in *mi*cro and *re*ciprocal and the initial letter in *degrees*.

Misery index
An index suggested in correspondence to *The Times*[38] as a classification of the discomfort endured during a wet English summer. In this index the balmy air and blue sky of the Mediterranean are indicated as 100, values below 30 call for immediate emigration to escape from the grey skies and continuous rains of Britain.

Mitotic index
The proportion of cells of living matter which are in mitosis (i.e. in nuclear division) at any given time.

Mohm
A unit of mechanical mobility equal to the reciprocal of the mechanical ohm, just as the mho is the reciprocal of the electrical ohm. It can be expressed in either CGS or MKS units and its dimensions are second mass^{-1}. The name was devised by Beranek[39] who derived it from *M*obile *ohm* in 1954.

Mohr cubic centimetre
A unit of volume used in saccharimetry[40]. It is the volume occupied by one gram of water at a specified temperature which is usually 17·5°C. In this instance the Mohr c.c. is equal to 1·00238 cm^3 or 1·00235 ml. The unit is named after the pharmacist C. F. Mohr (1806–79).

Mohs' scale
A hardness scale. (*See* **Hardness numbers.**)

Mol
See **Mole**.

Molal and molar solution
A molal solution contains one gram molecular weight of solute per thousand grams of solvent, whereas a molar solution contains one gram molecular weight of solute per litre of solution. W. F. Ostwald (1853–1932) was the first chemist to differentiate between a molal and a molar solution. The names molar and molal first appeared in English in 1902[41] and 1908[42] respectively.

Molality (m)
The molality of a solution is given by 1000 N where N is the number of moles of solute per unit mass (gram) of solvent[43]. The osmolality of a solution is

the molality an ideal solution of a non-dissociating substance must possess in order to exert the same osmotic pressure as the solution under consideration. It is often used in biology and medicine.

Molar fraction
A molar fraction is the ratio of the number of molecules or gram molecules of a specific constituent to the total number of molecules or gram molecules in the mixture.

Molarity
The molarity of a substance is the number of moles of solute per litre (cubic decimetre) of solution.

The osmolarity of a solution is the molarity an ideal solution of a non-dissociating substance must possess in order to exert the same osmotic pressure as the solution under consideration. It is often used in biology and medicine.

Mole (mol)
At the 14th CGPM (1972) the mole was approved as the unit of quantity of matter and was declared to be one of the seven base units of the SI system[44]. It is defined as the amount of substance of a system which contains as many elementary entities as there are atoms in 0·012 kg of carbon 12. When the mole is used, the elementary entities must be specified and may be atoms, molecules, ions, electrons, other particles or specified groups of such particles. The name mole first appeared in 1902[41] when it was used to express the gram molecular weight of a substance, so for example 1 mole of hydrochloric acid weighs 36·5 grams (atomic weights Cl 35·5, H 1). An abbreviated form, spelt mol, was suggested in 1948[45].

Moment
A unit of time. (*See* **Degré**.)

Mon
A unit proposed in 1964 for describing the flatness of rolled steel plates[46]. A surface has a flatness of 1 mon if no part of it is more than 25 μm above or below a straight line drawn between any two points 1 metre apart on the surface.

Month
The interval between successive new moons (approximately $29\frac{1}{4}$ days) has been in use as a unit of time since the dawn of history and the sub-division of the year into twelve calendar months, varying in length between 28 and 31 days, became customary throughout Europe at the time of the Roman

Empire. For scientific purposes the period between consecutive new moons is known as a synodic or lunar month (29·530 859 d); whereas a sidereal month, which is the time taken for the moon to complete one circuit of the earth with respect to the fixed stars, is 27·321 661 d compared with 27·321 582 d for a mean tropical month, a value which is derived from the observation of the movement of the moon between successive spring equinoxes.

Mooney unit
The plasticity of raw, or unvulcanized, rubber is sometimes given in terms of the torque on a disc situated in a cylindrical vessel containing rubber at a temperature of 100°C. The vessel rotates at 2 revolutions per minute and the torque on the disc after the vessel has been rotating for T minutes is measured on an arbitrary scale calibrated between 0 and 200. The number of the scale indicates the plasticity of the rubber in Mooney units[47], the result being expressed as so many Mooney units in T minutes. The unit is named after Dr Melvin Mooney, who devised the method in 1934[48].

Morgan
The unit of gene separation. It is the distance corresponding to unit probability of separation of two genes. Comparison with the physical dimensions of chromosome material leads to the conclusion that 10^{-5} morgan is about 30 Å and 1 morgan has been defined as the distance along the chromosome in a gene which gives a recombination frequency of 1 per cent. It is named after T. H. Morgan (1866–1945) the American Nobel Laureate of 1932 who was the author of the gene theory.

Moszkowski unit
A unit of transition probability used in nuclear physics. It was suggested[49] in 1955. The Moszkowski unit is of the same order of magnitude as the Weisskopf unit.

Mug
A name suggested in 1960 for the mass which is accelerated by one metre second $^{-2}$ by a force of kilogram weight[50]. The unit is called a Technische Mass Einheit in Germany – Engineering Mass Unit – which is abbreviated to TME. It is sometimes called a metric slug (q.v.).

Multiples and numbers
In April 1795 the French Revolutionary Government introduced prefixes to represent the multiples and submultiples of the basic metric units. Those given below, with their recognized modern abbreviations, are still in use.

10^3, kilo (k); 10^2, hecto (h); 10^1, deca (da); 10^{-1}, deci (d); 10^{-2}, centi (c); 10^{-3}, milli (m).

In the past hundred and fifty years the following additional prefixes have been added to the list:

10^{18}, exa (E); 10^{15}, peta (P); 10^{12}, tera (T); 10^9, giga (G); 10^6, mega (M); 10^{-6}, micro (μ); 10^{-9} nano (n); 10^{-12}, pico (p); 10^{-15}, femto (f); 10^{-18}, atto (a).

The 9th CGPM (1948) approved a nomenclature for large numbers[51] based on multiples of a million in which the name (N)illion is used to indicate values of 10^{6N}. Thus, when $N = 1$, the resulting number 10^6 is called a million; with $N = 2$ the number 10^{12} is a billion; with $N = 3$ it is 10^{18}, a trillion; $N = 4$, 10^{24}, a quadrillion. A different practice, however, is followed by financiers and many of the public on both sides of the Atlantic in which the nomenclature is based on $10^{(3Y+3)}$, thus when $Y = 1$, the number (10^6) is still a million, but 10^9 is called a billion, 10^{12} a trillion and 10^{15} a quadrillion. This system is not recommended by IUPAP[52].

The prefix micro micro was at one time used to indicate 10^{12} but it was superseded by pico during the 1939–45 war.

The 13th CGPM (1968) recommended[27] that powers of 10 should be expressed in units of $10^{\pm 3n}$ where n is an integer; thus the Planck constant should be written as $662 \cdot 6 \times 10^{-36}$ J s whereas it had previously been given as $6 \cdot 626 \times 10^{-34}$ J s.

In the 1860s G. J. Stoney[53] proposed a nomenclature in which powers of ten were indicated by an appended cardinal number when the exponent of 10 was positive and a prefixed ordinal when the exponent was negative; thus 10^{10} metre would be called a metre ten and 10^{-10} metre a tenth metre.

Musical scales[54]

The musical interval between two notes of frequencies f_1 and f_2 is defined as the ratio f_2/f_1. In every case the notes forming these intervals have a fixed and simple frequency ratio. The intervals most commonly used together with their names are given in Table 4. A musical scale is a series of notes ascending or descending in frequency by equal intervals suitable for musical purposes.

TABLE 4. *Musical Intervals*

Unison 1:1	Fourth 4:3
Semitone 16:15	Fifth 3:2
Minor tone 10:9	Minor sixth 8:5
Major tone 9:8	Major sixth 5:3
Minor third 6:5	Seventh 15:8
Major third 5:4	Octave 2:1

The octave is taken as the fundamental unit and this is subdivided into a number of smaller parts. Thus in the chromatic scale there are twelve equal semitones so that the ratio of the frequencies of each successive semitone is $2^{1/12}$. The twelve black and white notes on the keyboard of a piano are arranged in a chromatic scale. Other common scales are the scales of just intonation (diatonic scale) and of equal temperament. The latter scale first appeared in China and its establishment in Western music is usually ascribed to Bach. Each of these scales consists of eight notes. In the equal temperament major scale the frequencies are in the ratio 1, $2^{2/12}$, $2^{4/12}$, $2^{5/12}$, $2^{7/12}$, $2^{9/12}$, $2^{11/12}$ and 2. In the major scale of just intonation the ratios are 24, 27, 30, 32, 36, 40, 45 and 48. The major scales can be changed to minor scales by reducing the third and sixth notes by a semitone so that for example in the equal temperament minor scale the ratios $2^{4/12}$ and $2^{9/12}$ become $2^{3/12}$ and $2^{8/12}$ respectively. Since the notes in successive octaves form a geometrical progression with two as the common ratio, calculations may often be simplified by using a logarithmic scale. Thus if I be an interval measured in logarithmic units and f_1 and f_2 the frequencies of the notes forming an interval, then

$$I = K \log_{10} f_1/f_2$$

Taking the octave as the fundamental unit so that $I = K \log_{10} 2$, then K may be suitably chosen to give a reasonable number of units. If $K = 1000$, then the octave is divided into 301 units each of which is called a savart after the French physicist (F. Savart 1791–1841). This system came into use in about 1930. Sometimes the figure 300 is used but the difference between 300 and 301 is too small to cause inconvenience to musicians. If K is such that $I = 1200$, then the octave is divided into 1200 units which are called cents. This unit is used in America and Germany and was named by A. J. Ellis, the English translator of the book *Sensations of Tone* by H. von Helmholtz: it dates from 1895. If K be chosen to make $I = 100$, the resulting unit is the centi-octave. This unit corresponds to a frequency ratio given by $1 \cdot 007 : 1$.

The relationships between the musical interval, the savart, the cent and the centi-octave are given in Table 5.

TABLE 5

Frequency (Hz)	Ratio	Savart	Cent	Centi-octave
240	1	0	0	0
300	5:4	97	386	32·2
360	3:2	176	702	58·5
480	2:1	301	1200	100

N

n unit

A unit of neutron dose. It is the dose of fast neutrons which, when incident on an ionization chamber of specified characteristics, produces the same amount of ionization as would 1 roentgen of X-rays. The unit was in use by 1942[1] but is now obsolete.

Nanon

Spectroscopists sometimes call the length 10^{-9} m a nanon; the name is not recommended.

Neper (Np)

A number used mainly in continental Europe to express the ratio of two powers as a natural logarithm. Two powers P_1 and P_2 differ by n neper where $P_1/P_2 = e^{2n}$, i.e. $n = \frac{1}{2} \ln P_1/P_2$. The unit is called after John Napier (1550–1617) (Latin form Ioanne Nepero), the Scottish farmer-mathematician who prepared a table of the logarithms of trigonometrical functions in 1614[2]. The neper is equal to 8·686 decibels. The unit was approved in 1928, but by that time it had already been in use for four years[3].

Nepit

A unit of information defined as $I = \log_e P/P_0$, where P_0 and P are the probabilities at the receiver before and after reception of the message and I is the quantity of information expressed in nepits, a name derived from neper and digit. The unit is also called a nit.

New candle

A name adopted by the CIE in 1937[4] for the CGS and MKS unit of light intensity. It is synonymous with the candela (q.v.).

Newton (N)

The unit of force in the SI system of units. It is the force required to accelerate

1 kilogram by 1 metre per second per second and is equal to 10^5 dynes or about 100 grams weight. The unit was first proposed in 1900[5] when the name suggested was the large dyne. Four years later[6] the name newton (Sir Isaac Newton 1642–1727) was put forward but the name was seldom used until Hartshorn and Vigoureux[7] revived it in 1935. It was authorized by the IEC in 1938 and adopted by the 9th CGPM in 1948.

Nile
A unit of nuclear radioactivity sometimes used in Britain[8].

Nit (nt)
A unit of luminance in the MKS system which is equivalent to 1 candela per square metre or to 10^{-4} stilb. The luminance of the clear blue sky is within 2 to 6×10^3 nt. The unit came into use in 1948[9].

Noggin
A unit of volume equal to a quarter of a pint. The name was originally applied to a small drinking cup in the Gaelic-speaking parts of the British Isles but gradually became associated with the quantity of liquid (generally spirits) which such a cup could contain. In parts of the north of England the large noggin is used; there are two of these to the pint.

Noise rating number (Na)
This is an arbitrary number based on the fact that different amounts of noise can be tolerated under different conditions; thus a noise which would be acceptable in a workshop would be intolerable in an office. The noise rating number varies with frequency, but at 1000 Hz it is equal to the noise level expressed in decibels with respect to a noise level of 2×10^{-5} Nm^{-2}; this is the smallest pressure difference which can be detected by the average human ear. The noise rating numbers of a bedroom and a workshop are 25 and 65 respectively. The number was proposed by Rosenblith and Stevens in 1953 and is used extensively in the study of aircraft noise[10].

Normal solution
A normal solution contains 1 gram equivalent of replaceable hydrogen per litre of solution. Thus a normal solution of hydrochloric acid contains 1 gram molecular weight of hydrochloric acid (HCl) per litre of solution, and a normal solution of sulphuric acid (H_2SO_4) contains half a gram molecular weight of sulphuric acid. Normal solutions are often referred to as N solutions, e.g. an N/10 solution contains one tenth of the solute found in a normal solution.

The idea of normal solution dates from the early days of volumetric analysis which was begun by L. J. Gay-Lussac (1778–1850) in the third decade of the nineteenth century[11]. The name normal solution was in current use by the end of the nineteenth century[12].

Nox
A suggested unit for poor illumination. It is equivalent to 10^{-3} lux or 10^{-3} lumen per square metre. The unit was introduced in Germany[13] for measuring illumination during the 'black-out' in World War II.

Noy
A unit of perceived noisiness. One noy is defined as the perceived noisiness of the frequency band 910–1090 Hz of random noise at a sound pressure level of 40 dB above 0·0002 microbar. It is assumed that the perceived noisiness of a sound increases as a function of the physical intensity at the same rate as the loudness increases with intensity. The noisiness of a jet aircraft taking off is about 110 noys. The unit was proposed and named by Kryter[14] in 1959.

Nu value
This is a name given to the reciprocal of the dispersive power of transparent materials. The value is also called the constringence and is usually denoted by the Greek letter v, from which the name is derived. It is used mainly in the glass industry where Nu values are calculated from the equation $v = (n_D - 1)/(n_F - n_C)$, in which n_D, n_C and n_F are the refractive indices for the mean of the sodium D lines (589·3 nm) and the hydrogen red (656·3 nm) and blue (486·1 nm) lines respectively. Crown glass has Nu values ranging from 51 to 64 and flint glass from 21 to 55.

Number density (N)
The term number density is used to indicate the number of moles of substance per unit volume.

Numbers (Dirac): big number hypothesis
In 1937 Dirac pointed out[15 and 16] that the ratio of the largest to the smallest natural units of length, of force and of time, each came to about 10^{40}; a number which he called a cosmological constant; they have subsequently been often referred to as big numbers. Thus

$$\frac{\text{radius of the universe}}{\text{radius of an electron}} \simeq \frac{\text{coulomb force between proton and electron}}{\text{gravitational force between proton and electron}}$$
$$\simeq \frac{T}{t}$$
$$\simeq 10^{40}$$

where T is the time taken for light to reach the edge of the universe and t that required for light to cover a distance equal to the radius of an electron.

Furthermore, the ratio of the mass of the universe to the mass of an average particle of matter comes to about 10^{79}, which is approximately $(10^{40})^2$; the figure 10^{79} is also called the Eddington number after the British astronomer Sir Arthur Eddington (1882–1944) who predicted the number of particles in the universe in the late 1920s. The big number hypothesis indicates that if the fundamental constants are all expressed as dimensionless ratios to rid them of man-made dimensions (e.g. by using units of length, mass and time from the hydrogen atom) the constants will tend to lie within a few orders of magnitude of unity with the notable exception of the gravitational constant G.

Nusselt number (Nu)
An alternative name for the Biot number (q.v.).

O

Octane number
A number used in automobile engineering to describe the 'anti-knock' properties of spark ignition fuels. (*See* **Cetane number**.)

Octet
See **Byte**.

Oersted (Oe)
The oersted is the unit of magnetizing force, or magnetic field strength, in the CGS system of units; it is usually denoted by H. The unit may also be expressed in abamperes per centimetre and is related to magnetic flux density, B, by the relationship $B = \mu H$, where μ is the permeability of the medium. When air is the medium $\mu = 1$, so that it is not then necessary to distinguish between B and H when making CGS magnetic calculations involving this medium.

The unit was defined at the Oslo meeting of the IEC in 1930 when it was named after the Danish (Scandinavian) physicist[1] H. C. Oersted (1777–1851) the first person to investigate the relationship between electric current and its magnetic field. At one time the unit of magnetizing force was called the gauss[2] and, in 1903, the name oersted was suggested by some workers in the USA for the CGS unit of reluctance[3], but the proposal was not accepted.

Oerstedt
In Everett's book *Units and Physical Constants*, which was first published[4] in 1879, there appears the phrase: 'The unit of current in the Indian Telegraph Department is called the Oerstedt'.

Ohm (Ω)
The ohm is the SI unit of resistance. The unit of resistance is the oldest of the

electrical units; it dates back to 1838[5] when Lenz produced a standard resistor which consisted of a foot of No. 11 copper wire. In the next twenty-five years two general types of resistance standard were gradually evolved, one for the laboratory and the other for the outdoor telegraph engineer. The former consisted of short lengths (less than a metre) of a specified conductor and the latter were long lengths (a mile or kilometre) of telegraph wire. The first attempt to get order out of chaos took place in 1861 when the British Association set up a committee to report on electrical units. This committee proposed an absolute system based initially on the millimetre, the gram and the second. A later committee in 1873[6] favoured the centimetre, the gram and the second as the fundamental units. Both committees, however, suggested that practical units could be obtained by multiplying the fundamental units by a suitable power of ten to make them a convenient size for practical work. Thus the fundamental unit of resistance in the CGS system could be multiplied by 10^9 to make it have the same value as the resistance of a siemens unit, which was a column of mercury 1 metre long and 1 square millimetre in cross-section. This new practical unit was named the ohm after G. S. Ohm (1787–1854) who discovered the relationship between electrical current, resistance and potential difference in 1826. Some ten years before this, however, Sir Charles Bright and Latimer Clark, the leading telegraph engineers of their day, had suggested that the unit of resistance should be called a volt[7]; another name suggested was the ohmad.

The ohm was established internationally at the first meeting of the IEC[8] in 1881. This meeting defined the legal ohm in absolute terms but stated that for practical purposes its resistance was that of a column of mercury 106 cm long, 1 mm^2 in cross-sectional area, at a temperature of 0°C. Within a few years it was recognized the resistance was too low, and a new ohm in which the length of the mercury column was increased to 106·3 cm was defined in 1891. This new ohm was ratified by the Board of Trade shortly afterwards and it remained the standard of resistance until replaced by the International ohm in 1908[9]. The 1908 meeting defined the ohm as the resistance of a column of mercury 106·300 cm long and 14·4521 g in mass at the temperature of melting ice. The same meeting agreed that the ohm was to be one of the two primary electrical standards; the other was the ampere. The international ohm remained the standard until replaced by the absolute ohm in 1948[10]. The change was agreed in 1935 when it was apparent that not only was the international ohm more difficult to reproduce accurately than the absolute ohm but the international ohm was also too small[11]. The SI, or absolute, ohm is now defined in terms of work, current and time. The abbreviations, ω and Ω used for the ohm were suggested by William Preece when he was acting as an instructor to Indian Telegraph Service cadets at the Hartley College (now the University of Southampton), in 1867[12]. The various values of the unit of resistance are given in Table 6.

TABLE 6. *Values of unit of resistance*

Date	Name	Type of resistor	Approximate value absolute ohms
1838	Lenz	1 ft of No. 11 copper wire (diam. 0·1 in)	8×10^{-4}
1843	Wheatstone	1 ft of copper wire weighing 100 grains	0·025
1848	Jacobi	25 ft of copper wire weighing 345 grains	0·64
1850	English†	1 mile No. 16 impure copper wire (diam. 1/16 in.)	25
1850	French†	1 km of iron wire 4 mm in diameter	10
1850	German†	1 German mile (8238 yds) of No. 8 iron wire (diam. 1/6 in.)	55
1860	Siemens	1 metre of mercury of 1 mm² cross-section at 0°C	0·96
1865	B.A. ohm	'platinum–silver alloy replica of the absolute unit'	0·9866
1883	Legal ohm	106 cm of mercury, 1 mm² cross-section at 0°C.	0·9972
1894	Board of trade ohm	106·3 cm of mercury of mass 14·4521 g at 0°C.	1·000
1908	International ohm	106·300 cm of mercury of mass 14·4521 g at 0°C.	1·000*
1948	Absolute ohm		1·000

* 1 international ohm = 1·00049 absolute ohms.

† Telegraph Engineers Units (temperature not specified).

Ohma
An early name for the practical unit of potential. It was suggested by Bright and Clark[7] in 1861.

Ohmad
A name at one time used for the practical unit of resistance. It was proposed in 1865[13] and was replaced by the name ohm in 1881.

Open window unit (o.w.u.)
A unit of sound absorption first used by Wallace C. Sabine[14] in 1911. It is equal to the absorption by an open window of 1 square foot in area. The unit was renamed the sabin in 1937.

Optical density
This is a number which indicates the light transmission characteristic of a material. If the ratio of the intensity of transmitted light to the incident light is T, then the optical density D is given by $D = \log_{10} 1/T$. Since $T = e^{-\alpha t}$, where α is the absorption coefficient and t is the thickness of the material, $D = \log_{10} e^{\alpha t}$. Thus if α is known the optical density corresponding to a certain thickness can be found. A filter classified as having an optical density of 2 means that the light transmitted is reduced in intensity by 100. Optical density was first

employed by Hurter and Driffield[15]. Sometimes, particularly for liquids, T is given by $T = 10^{-atc}$, where a is the absorptivity or extinction coefficient and c is the concentration. As atc is dimensionless, the units of a are the reciprocal of cell length times concentration, and are expressed as $cm^2\, mole^{-1}$ or $cm^2\, g^{-1}$.

Osmole (osm)
The unit of osmolality and osmolarity. (*see* **Molality** and **Molarity**).

Ounce
A unit of mass in both the Troy and the avoirdupois systems. In the former, where it dates from the tenth century, it is equal to $\frac{1}{12}$ pound (Troy) whereas in the avoirdupois system there are 16 ounces in the pound.

P

Paper sizes

The sizes to which trimmed paper and board are cut after manufacture are indicated by the letters A, B and C[1]. The basic areas of these three sizes are 1, $2^{1/2}$ and $2^{1/4}$ square metres respectively. If y be the length and x the breadth of a basic sheet, x and y are so chosen that $y = x\sqrt{2}$, hence the basic length of a sheet in the A is 1189 mm and its breadth 841 mm and such a sheet is described as A0. If it now be folded by dividing its y axis into two equal parts, its dimensions become 841 mm and 594 mm and it is described as A1. Size A2 is obtained by again folding the longer axis in two thereby making y and x 594 mm and 420 mm respectively. Smaller sizes can be obtained by continuing this folding, see Table 7. Sizes in the B and C ranges are obtained in a similar manner by folding the appropriate basic sheets; thus size B0 has $y = 1414$ mm and $x = 1000$ mm, whereas C0 has $y = 1297$ mm and $x = 917$ mm. When the designation or code letter is preceded by a fraction, it indicates that the basic length has been reduced by the fraction concerned; e.g. $\frac{1}{4}$A0 has a length of 841 mm and a breadth of 297 mm, the latter being a quarter of the original length of the A0 sheet. The dimensions for $\frac{1}{4}$A1 are 594 mm and 210 mm. The prefixes R and SR in front of the designation letter

TABLE 7. *International paper sizes*

Code	x (mm)	y (mm)	Code	x (mm)	y (mm)	Code	x (mm)	y (mm)
A0	841	1189	B0	1000	1414	C0	917	1297
A1	594	841	B1	707	1000	C1	648	917
A2	420	594	B2	500	707	C2	458	648
A3	297	420	B3	353	500	C3	324	458
A4	210	297	B4	250	353	C4	229	324
A5	148	210	B5	176	250	C5	162	229
A6	105	148	B6	125	176	C6	114	162

show that the sheet is untrimmed but will be reduced to the standard indicated on trimming, thus RA0 and SRA0 reduce to A0 after cutting.

Par

A name suggested in 1940[2] for the metric slug. It is derived from the first three letters of the french adjective *paresseux* – sluggish. (*See* **Metric slug**.)

Parker

See **Roentgen equivalent physical**.

Parsec (pc)

A unit used by astronomers to describe stellar distance. It is the distance from which the earth's orbit would subtend an angle of one second of arc, i.e. 308.6×10^{14} m, 3.26 light years or 206×10^3 astronomical units. The unit was approved at the first meeting of the International Astronomical Union which took place in 1922[3]. The name, derived from *par*allax *sec*ond, is said to have been proposed by Professor H. H. Turner (1861–1930), Savilian Professor of Astronomy at Oxford. The idea of using star parallaxes for measuring astronomical distances was suggested about 1840 by both Bessel and Meadows.

Particle sizes

Geologists classify the sizes of particles on several scales, the one in most general use today being that proposed by Udden in 1898 and modified by Wentworth in 1922[4]. In this, particles less than 1/256 mm in diameter are classified as clays, those with diameters in the range 1/256 to 1/16 mm are known as silts, 1/16 to 2 mm are sands, 2–64 mm pebbles, 64–256 mm cobbles and those with diameters greater than 256 mm are called boulders. Other scales include those of the US Bureau of Soils, Crook (1913), Boswell (1918) and Hatch and Rastall (1923). In all these scales the same general classification is followed but the dividing lines between each group vary slightly from scale to scale; thus the maximum diameter of sands is 2 mm in the Boswell scale, 1·5 mm in the Hatch and Rastall scales and 1 mm in the Crook scale. In many of the scales the sands are further sub-divided into coarse, medium, fine and very fine.

Pascal (Pa)

The unit of stress in SI units. It is equal to a force of 1 newton per square metre so that 1 pascal is equivalent to 10 microbars. The unit is named after B. Pascal (1623–62) who was the first man to use a barometer for measuring altitude. The name appeared in the tables of Kaye and Laby in 1956[5] and the unit was officially approved at the 14th CGPM in 1971.

Pascal second
The SI unit of viscosity.

Pastille dose
The dose of radiation required to change the colour of a basic platinocyanide pastille from apple green (Tint A) to reddish brown (Tint B). The unit is now obsolete[6]. It was equal to about 500 roentgens and was originally called a B dose when it was introduced after the 1914–18 war.

Peck
A unit of capacity equal to 2 gallons. It is used in both liquid and dry measure and dates from the early fourteenth century.

Peclet number (Pe)
The Peclet number plays the same role in the transfer of thermal energy in fluids as does the Reynolds number in the transfer of momentum. It is equal to the product of Prandtl and Reynolds numbers and is given by $\rho cvl/k$ or vl/K, where ρ is the density, c is the specific heat, v is the velocity, l has the dimensions of length, k is thermal conductivity and K is the thermal diffusivity or thermometric conductivity. The name was proposed by Grober[7] in 1921 and commemorates J. C. E. Peclet (1793–1857) who was the first to apply Fourier's ideas of thermal conductivity to engineering problems. The number was recognized[8] in 1941 by the American Standards Association.

The Peclet number (Pe*) used in mass transfer problems is given by vl/D where D is the diffusion coefficient.

Pencil hardness
The letters H, B and F – indicating hard, black and firm – which are used to indicate the hardness of the lead in pencils are reputed to have been introduced by the German firm of J. J. Rehbach in 1823[9]. The numbers defining the degree of hardness (or blackness) are a later refinement. In recent years the designation F has been replaced by HB.

Perch
A unit of length equal to $16\frac{1}{2}$ feet. The unit has been in use in England from about the time of Henry II (1154–1189). It is called a pole or a rod. In Ireland the perch used to be 21 feet, which increased the length of the Irish mile to 2240 yards.

Permicron
A name proposed[10] in 1951 for the number of wavelengths in a micron (now micrometre) so that if the wavelength of a spectral line is 540 nm its

wave number would be 1·85 permicrons. Its proposer claimed that the permicron would be more convenient than the kayser (q.v.) when the wavelength concerned could not be measured to an accuracy better than 1 in 1000.

pH index

The pH index is a number used to express the hydrogen ion concentration of a solution. The conception of pH is due to Sorensen[11] who in 1909 suggested that numerical values should be used as the basis for an acidity scale. He used the symbol P_H^+ for his scale but this was subsequently abbreviated to pH. The pH index is defined as the common logarithm, with sign reversed, of the hydrogen ion concentration expressed in gram ions per litre of solution and it lies between 0 and 14. Acids have a value of less than 7, alkalis have an index greater than 7, and 7 represents the index for a neutral solution[12]. Values for pH can be determined electrically or by chemical indicators, the latter giving sufficient accuracy for normal laboratory work.

Other scales have been proposed for expressing the acidity of a solution. These include the pR unit of Giribaldo (1938), the N_n unit of Gestle (1938), sometimes called the hydron, the rA number of Catani, the paH unit (1924), the pwH unit (1930) and the ptH unit (1938). The last three units are of the same order of magnitude as the pH index, but in the pR and N_n units alkalis have values between 0 and 14, whereas acids lie in the range 0 to −14 and zero represents a neutral solution[13].

Phon (P)

A unit of loudness which is a measure with reference to the sound pressure level of a pure tone of frequency 1000 Hz judged by the listener to be of equal loudness. If the sound pressure level be n decibels above the reference level of $20 \times 10^{-6}\,\mathrm{Nm^{-2}}$, the loudness is equal to n phon. The loudness of a jet aircraft engine is about 140 phon, whereas the noise of a steam railway locomotive is less than 100 phon. The idea of using a reference frequency of 1000 Hz for loudness was suggested in 1927[14]. The phon as at present defined[15] is of German origin, but when originally proposed in 1933[16] it had a reference sound pressure level of $31\cdot6 \times 10^{-5}\,\mathrm{Nm^{-2}}$. It was one of the two acoustical units defined at the first International Acoustical Conference held in Paris in 1937, the other unit being the decibel[17]. For the relationship between the phon and the unit of subjective loudness (*see* **Sone**).

Phot

The unit of illumination of a surface in the CGS system. It is equal to 1 lumen per square centimetre. It is somewhat large and most illuminations are more conveniently described in milliphots.

The unit was legalized in France[18] in 1919 and two years later it was named the phot by A. Blondel[19].

The phot-sec is a CGS unit for exposure. It is recognized by the CIE.

Photo-absorption unit

In photoelectric calorimetry the photo-absorption of a solution is expressed in terms of the mass in grams per millimetre of solution required to reduce the light transmitted through one centimetre of solution by 1 per cent. This may be written[20] either as $E_{1\,cm}^{\%}$ or E(1%, 1 cm). In 1947[21] a proposal was made to use the number of grams per millilitre needed to reduce the transmitted light by 50 per cent and this unit was designated $W_{1\,cm}^{50}$ or W(50, 1 cm). The relationship between E(1%, 1 cm) and W(50, 1 cm) is therefore E(1%, 1 cm) × W(50, 1 cm) = 0·00301, provided the Beer–Lambert law holds for the solution concerned.

Photographic emulsion speed indicators

These indicators give the response of photographic emulsion to light. The systems in use today are the H and D, the Scheiner, the DIN, the Weston, the Ilford, the BSI and the ASA. The earliest method of giving emulsion speed was described by Warnerke in about 1879, but the first usable system was the H and D proposed in 1891[22] by F. Hurter (1844–98) and V. C. Driffield (1848–1915). They established a characteristic curve which was a graph of density (*D*) against the relative logarithm of the exposure time (log *E*). The central part of this curve is linear and from the point where the extrapolated line cuts the log *E* axis the H and D speed number is found. This indicates the speed of an emulsion at its maximum sensitivity and the numbers get greater as the speed increases. The slope of the linear part of the curve is called the gamma[23] (q.v.). The Weston number, the British Standards Institution speed number and the American Standards Association system also indicate the speed of the emulsion at its maximum sensitivity.

The Scheiner and the DIN (Deutsche Industrie Norm) systems give a series of numbers, called degrees, based on the minimum exposure required to get a detectable image formed on the emulsion. These systems express the performance of the emulsion on a logarithmic scale, whereas all the other systems mentioned use an arithmetic scale. The Scheiner scale was devised[24] in 1898 by J. Scheiner (1858–1913) at Vienna and was first used commercially by the Secco Film Company of Boston, Mass. in 1899. The other scales are of more recent origin. The Ilford scale is unique in that it uses letters to indicate the performance of an emulsion. A comparison of the systems is given in Table 8[25].

Photon

The name given to a packet of radiant energy of value *hv* joules or

TABLE 8. *Working correlations between speed systems*

Relative speed	1	4	16	64	128
H and D	100	400	1600	6400	12800
Scheiner BSI	17°	23°	29°	35°	38°
Scheiner American	12°	18°	24°	30°	33°
DIN/10	7°	13°	19°	25°	28°
Weston II	2·5	10	40	160	320
ASA and Weston III	4	16	64	250	500
BSI	016	064	0252	01000	02000
Ilford	A	C	E	G	H

$hv/(160 \times 10^{-21})$ electron volts, where h is Planck's constant and v is the frequency of the radiation. (*See* **Quantum**.)

Pica
A printer's unit of type width. (*See* **Point (printers')**.)

Pieze (pz)
The pieze is the basic unit of pressure in the metre tonne second system and is equal to 1 sthéne per square metre or 10^3 Nm^{-2}. The pieze received legal recognition in France[18] in 1919.

Pint
A measure of capacity equal to $\frac{1}{8}$ gallon. In Great Britain it is equal to 0·57 litre, in America to 0·47 litre. The unit dates from the fourteenth century.

Planck
A unit of action equal to one joule in one second, i.e. it has the dimensions of Planck's constant. The unit was proposed in 1946 and is named after Max Planck (1858–1947) the originator of the quantum theory[26].

A proposal was made in 1972 to call the SI unit of angular momentum a planck[27], but it has met with little support[28].

Planck constant (*h*)
See Appendix 1.

Point (jewellers')
A unit of mass equal to 0·1 carat which is used in the weighing of diamonds. (*See* **Carat**.)

Point (printers')
The depth of type face is measured in points where 1 point is approximately 1/72 inch. Type sizes are also indicated by names, e.g. brilliant (3·5 points), emerald (6·5) and pica (12). The term pica is also used to describe type width, there being 6 picas to an inch.

Points of the compass
When a complete circle representing 360 degrees is divided into 32 units, each unit represents an angle of $11\frac{1}{4}$ degrees and is called a point of the compass. The 32 points of the compass were in use by the mid-thirteenth century when the seaman's horizon was divided into 32 parts called the 'rhumbs of the wind'.

Poise (P)
The poise is the CGS unit of dynamic viscosity. It is named after J. L. M. Poiseuille (1799–1869) one of the early workers on capillary flow. The unit has the dimensions g cm^{-1} sec^{-1}, but in practice the centipoise (cP) is of a more convenient size than the poise for describing the viscosity of freely flowing liquids, e.g. acetone 0·324 cP at 20°C. The viscosity of gases is frequently expressed in micropoises (μP); (air 181 μP at 20°C). The British Standards Institution[29] recommends that for calibrating viscometers the dynamic viscosity of water at 20°C should be taken as 1·0020 cP.

The poise was suggested[30] as the unit of viscosity in 1913, but some authorities considered the centipoise should have been the unit[31], as its size was more convenient.

Pole
A unit of length equal to $16\frac{1}{2}$ feet. The unit is also called a perch or a rod. The name pole came into use in the early sixteenth century.

Pole strength
There is no name for the unit of pole strength in the CGS units. The strength of a pole is given in terms of the force in dynes acting on it when placed in a field of strength H; thus a force of mH dynes indicates a pole strength of m. The magnetic induction of such a pole is $4\pi m$ maxwell.

In the rationalized MKS units, if Φ weber be the total flux associated with a magnetic pole, its pole strength is Φ weber. In the nineties pole strengths were sometimes expressed in a unit called the weber[32]. A pole had a pole strength of 1 weber when it produced a field of 1 oersted at a distance of 1 centimetre in air.

Pollen count

A term used by botanists to indicate the number of grains of pollen of all kinds per unit area of microscope slide; in practice the area is often taken as 1 square centimetre. The term came into use in the early 1930s[33].

Poncelet

A metric unit of power. It is the work done in 1 second by a force acting over a metre, the force being able to accelerate 100 kg by a metre per second per second. It was legalized in France[18] in 1919 and is named after J. V. Poncelet (1788–1867) an early worker on pressure; the unit is equal to 980·665 watts.

Pond

A noncoherent unit of force equal to one gram weight.

Postages (P)

Between about 1850 and 1890 postal weights were calculated in half ounce units known as postages, a postage being 1 penny per half-ounce. Thus since 4 ounces equals 8 half-ounces it was written as 8P or P/8 and similarly 2 ounces would be written 4P or P/4. Postal weights marked F referred to foreign postages which were 1 penny per quarter-ounce, thus 2F = 1P.

Poumar

A unit used in the textile industry for describing the mass per unit length of a yarn. (*See* **Yarn counts**.)

Pound (lb)

The pound is the unit of mass in the foot pound second system of units. It is very old and can be traced back to the Roman Libra (0·327 kg). It has been defined in Great Britain on numerous occasions from the reign of Ethelred the Unready (968–1016) until the present day. From the time of Elizabeth I until quite recently two different pounds were recognized in the British Isles – the avoirdupois (7000 grains) and the troy (5760 grains). The latter was restricted to the weighing of gold, silver and precious stones by Parliament in 1834.

The Imperial Standard Pound as defined in the Weights and Measures Acts of 1855 and 1878 was, until 1963, the fundamental unit of mass in Great Britain. These Acts defined the pound as the mass of certain platinum weight of given dimensions. In 1963 the pound was defined as being equal to '0·45359237 kilogram exactly' by the Weights and Measures Act passed in that year[34]. This pound is identical with the international pound adopted in 1959[35] by Standards Laboratories throughout the world. Its value in kilogram is approximately equal to the mean of the Imperial Standard Pound[36] of 1878 (0·453592338 kg) and the pound used in the USA (0·4535924277 kg). The figure of 0·45359237 was deliberately chosen as a number divisible by seven to facilitate the conversion of grains to grams, there being 7000 grains in a pound and hence 1 grain = 0·06479891 gram.

Poundal (pl)
The poundal is the force which can accelerate a mass of 1 pound by a foot per second per second, i.e. 1 pound weight is equal to 32 poundals. 1 poundal is equal to 0·138255 newton. The name of the unit was suggested in 1876 by James Thomson[37], the brother of Lord Kelvin. Some elementary text books on dynamics used in schools use the corresponding forces for the ounce and the ton, namely ouncedal and tondal, to represent the force required to accelerate 1 ounce, or 1 ton, by 1 foot per second per second.

Power factor
The power factor is the ratio of the true watts to the apparent watts developed in a component of a circuit through which an alternating current is passing. It is equal to the cosine of the phase angle. The name power factor was introduced and defined by Fleming[38] in 1892.

Pragilbert
A pragilbert is a unit of magnetomotive force equal to 4π ampere turns[39].

Praoersted
A praoersted[39] is a unit of magnetizing force equal to 4π ampere turns metre^{-1}.

Prandtl number (Pr)
A dimensionless parameter used in fluid mechanics. It may be expressed as the ratio of the kinematic viscosity v to the thermal diffusivity or thermometric conductivity K. The latter is given by $k/\rho c$ where k is the thermal conductivity, ρ is the density and k is the specific heat. At 20°C the Prandtl number lies between 0·67 and 1·0 for gases; for water it is 6·7 and it is of the order of thousands for very viscous liquids. The unit was deduced by Nusselt[40] in 1910, but has been named after L. Prandtl (1875–1953) in the misinformed belief that he had initially derived it in 1922. Prandtl, in a footnote in his book[41] on fluid dynamics, states 'The author is unwilling to encourage this piece of historic inaccuracy and therefore prefers the equally brief expression v/K'. The number was recognized by the American Standards Association in 1941[8].

Preece

The preece was the name suggested for a unit of resistivity equal to 1 megohm quadrant – i.e. it was the same numerically as the resistance measured in megohms of a cube with sides 10^7 metres in length. It was used occasionally at the turn of the century to describe the resistivity of insulators; thus paraffin wax has a resistivity of about 34 preeces. The unit was suggested in 1891 by a former British Post Office Engineer-in-Chief[42], Sir William Preece (1834–1913), and was named after him by Professor Everett of Belfast some ten years later.

Predicted mean vote index

An index used to quantify the degree of discomfort when the optimal thermal environment cannot be achieved[43]. It gives the predicted mean vote of a large group of subjects according to the following psycho-physical scale: $+3$ hot, $+2$ warm, $+1$ slightly warm, 0 neutral, -1 slightly cool, -2 cool, -3 cold. PMV values are determined from tables. An acceptable thermal environment has a value of -0.5.

Predicted percentage of dissatisfied index

An index derived from the predicted mean vote index which gives an expression for the quality of a given thermal environment, i.e. it predicts the percentage of a large group of people likely to feel thermally uncomfortable.

Preferred numbers

Preferred numbers are conventionally rounded-off terms in a geometrical progression whereby a tenth multiple of the initial term is obtained after a predetermined number of terms, viz: $p, pq, pq^2, \ldots, pq^{n-1}, pq^n$, where p is the initial number, n is the number of terms and q a factor such that $q^n = 10$. The advantage of this system is that it can reduce the number of individual components of specified tolerance which is required to cover a given range of values. Thus in the case of resistors with a tolerance of 20 per cent, only seven components would be needed for the range R to $10R$ and these are R, $1.5R$, $2.2R$, $3.3R$, $4.7R$, $6.8R$ and $10R$. In this particular set of preferred values each component is $10^{1/6}$ larger than its predecessor and this series is called an $E6$ series. Other series of preferred values are available for components with tolerances of 5 per cent, 10 per cent, etc.[44].

Preferred numbers were first used by Captain Charles Renard, an engineer officer, when making artillery balloons in France[45] in the period 1877–79. Preferred numbers were standardized in Germany in 1920 and in France a year later. In 1927 a set of preferred numbers were informally approved by the American Standards Association and these received international approval[46] in 1935. The insatiable demand for electronic components in the 1939–45 war hastened the adoption of preferred values in the electrical industry. In 1953 the International Standards Organization approved four series of basic preferred numbers[47], which are known as the $R5$, $R10$, $R20$ and $R40$ series. In these series each consecutive number differs from its predecessor by factors of $10^{1/5}$, $10^{1/10}$, $10^{1/20}$, and $10^{1/40}$ respectively.

Priest

A name suggested in 1946 for the Z stimulus in the trichromatic colour system. (*See* **Trichromatic unit**.)

Prism dioptre

A unit for describing P, the deviating power of a prism. If θ be the angle of deviation, then $P = 100 \tan \theta$.

Promaxwell

A CGS unit of magnetic flux equal to 10^8 maxwells. It was proposed in 1930[48] but before it became widely used it was replaced by the weber. (1 weber = 10^8 maxwells.)

Prout

A unit of nuclear binding energy equal to one twelfth of the binding energy of the deuteron. Its value is 0·185 MeV or 195×10^{-6} amu. The unit was suggested by Witmer[49] in 1947, because the binding energies of most nuclei are frequently equal to some integral of this value. Heavy nuclei have binding energies of the order of 42 prouts, but the energies of light nuclei can be greater than this. The unit is named after the Scottish physicist William Prout (1786–1850) who put forward a theory that all atoms were composed of hydrogen atoms. Neither the name nor the unit has been widely adopted.

Puff

A name sometimes used to indicate a capacitance of 10^{-12} farad. It originated during the 1939–45 war and is derived from the pronunciation of 'pF' very quickly.

Pulsatance

A name suggested[50] in 1919 for angular frequency. It is defined as $2\pi f$, where f is the frequency in hertz. It is now obsolescent.

Pyron

This is a unit which has replaced the langley (q.v.) as the unit of solar radiation density. it is derived from the Greek word for fire.

Q

Q factor

A factor which describes the quality or selectivity of a circuit. It is numerically equal to $2\pi fL/R$ or $(LC)^{1/2}/RC$, where f is the resonant frequency of the circuit and L, C and R are the inductance, capacitance and resistance respectively. The letter Q was first used for the quality of a circuit in 1931[1] and by 1938 it was sufficiently well known to be included in *The Admiralty Handbook of Wireless Telegraphy*. In the early 1930s the letter m was sometimes employed to describe $2\pi fL/R$; in this case m was called the magnification of the circuit[2]. At about this time in Germany, d, the dissipation factor was used, this being equivalent to $1/Q$.

Q unit

A unit used originally by Sir John Cockcroft (1887–1967) in 1953 to express the world's fuel reserves[3]. $1\ Q = 10^{18}$ Btu. At that time the coal reserves were considered to be 33 Q.

Q-value

This is a measure of the energy that is released or absorbed in a nuclear reaction. It is sometimes called the disintegration energy. The Q-value may be either positive or negative. If it is positive then rest mass-energy is converted into kinetic or radiation mass-energy or both. If it is negative then either kinetic or radiation mass-energy must be supplied for the reactions to take place. When a Q-value is calculated, the masses used must be the rest masses of the nuclei and if the rest masses are in amu (q.v.), then the Q-value is also in amu.

Quad

This is the name given to 10^{15} Btu or 1.055×10^{18} J. The name[4] is derived from 10^{18} which is the American quadrillion and some users take the quad as being 10^{18} Btu.

113

Quadrant

A name used between 1889 and 1893[5] for the unit of inductance which today is called the henry (q.v.). It was also sometimes used at the end of the nineteenth century as a unit of length equal to 10^7 m, which is the length of an earth's quadrant.

Quantum

This is the name given to a packet of radiant energy of value hv joules or $hv/(160 \times 10^{-21})$ electron volts, where h is Planck's constant and v is the frequency of the radiation. This unit was introduced by Planck in 1900. An alternative name usually used when light energy is being considered is photon (*see* **Ergon**). The energy in a quantum of radiation having a wavelength 560 nm is about 3.55×10^{-19} joule so that a 100 watt bulb which emits 5 per cent of its energy as visible light gives approximately 1.4×10^{19} visible quanta per second.

Quart

A unit of volume in the English system which is used in both dry and liquid measures. There is a mention of a quart as a measure of ale in Chaucer's *Miller's Tale* (1390). One imperial quart = 1·13650 litre and there are 4 quarts in an Imperial gallon. In the English system there is no difference between the liquid and the dry quart; thus a quart of corn occupies the same volume as a quart of water. In the United States the liquid quart is 0·94633 litre and the dry quart 1·1012 litre.

Until the passing of the 1952 Customs and Excise Act, wines and spirits were measured in reputed quarts in the United Kingdom. A reputed quart consisted of 26·67 fluid ounces, i.e. 0·67 of an imperial gallon or 761×10^{-6} m^3.

The Winchester quart was one of the standards of volume used in England from medieval times until the passing of the 1835 Weights and Measures Act. It is still sometimes used today, viz. a Winchester quart of acid. One of the original standards is available for public inspection in Winchester. 1 Winchester quart $\simeq$ 0·99 imperial quarts.

Quevenne Scale

A scale used in lactometry. (*See* **Degree (hydrometry)**.)

Quintal (q)

A unit of mass used in England in the fifteenth century. It was equal to about 100 pounds. A metric quintal is equal to 100 kilograms (220·5 pounds).

R

R units (Solomon and German)

These were X-ray intensity units which were used in the period 1920 to 1930[1]. X-rays had an intensity of 1 Solomon R unit if they produced the same ionization as that from 1 gram of radium placed 2 cm from an ionization chamber when there was a platinum screen 0·5 mm thick between the source and the chamber. 1 R unit = 2100 r h^{-1}; 1 r s^{-1} = 1·71 R.

The German R unit was in use at the time as the Solomon unit and was equal to about 2·5 Solomon units.

R-value

This is a recent unit for insulation. It is a measure of thermal resistance (1 SI unit = 1 K m^{2}W^{-1}) and is the reciprocal of the U factor value (q.v.). British Standard 5803 specifies that the R value must be stated on the packaging of blanket materials. The larger the R value the better is the material as an insulator.

Rad

The unit of absorbed dose of radiation. The absorbed dose is 1 rad when 1 kilogram absorbs 10^{-2} joules of energy. The rad was defined in CGS units by the International Commission on Radiology[2] in 1953 and restated in SI units in 1970. In 1956[3] the rad replaced the roentgen (unit of radiation exposure) for clinical work involving X-rays or radioactive sources. In practice the rad and the roentgen both represent about the same amount of energy since 1000 rads equal 1100 ± 50 roentgens but, unlike the roentgen, the rad is applicable to all types of radiation. The exact equivalent depends on the material exposed. Historically the rad can be traced back to 1918. In this year Russ suggested that the unit of X-ray dose should be the dose required to kill a mouse and that the unit should be named the rad[4].

Radian (rad)

A unit of angular measure equal to the angle subtended at the centre of a circle by an arc the length of which is equal to the radius. The name is reputed to have been made up by James Thomson in an examination paper in Belfast about 1870 and it was first published in 1873[5]. There are 2π radians in a circle, i.e. one radian is equal to $57 \cdot 29578°$. The angle $\pi/4$ is called an octant and a thousandth of a radian is sometimes called a mil, which is equal to 3 minutes 26·5 seconds of arc.

Radian frequency
A name suggested for angular frequency in 1949[6]. It is given by $2\pi f$, where f is the frequency in hertz.

Radiation unit (ru)
A unit used in the measurement of the absorption of cosmic rays[7]. (*See* **Shower unit.**)

Ram or ray
Names suggested by McLachlan[8] in 1934 for the units of mechanical and acoustical impedances. The names were intended to commemorate the third Lord Rayleigh (1842–1919), a pioneer of acoustic measurement. (*See* **Acoustic ohm** and **Mechanical ohm.**)

Rankine scale
See **Temperature scale.**

Rapid plasticity number
A number used in characterizing viscosity for raw rubber and unvulcanized rubber mixes. A standard test consists of compressing a small disc of rubber (0·4 cm^3 in volume) between two heated platens at 100°C, first to a thickness of 1·00 mm where it is preheated for 15 s and then under a constant force of 100 N for a further 15 s. Then the thickness of the compressed test-piece is measured in units of 0.01 mm. The median of three readings is known as the rapid plasticity number. It is sometimes called the Wallace plasticity number.

Rayl
A unit of specific acoustic impedance. The specific acoustic impedance at a surface is defined as the ratio of the effective sound pressure to the effective particle velocity at the surface. This ratio is a complex number. The specific, or unit area, acoustic impedance is 1 rayl if unit effective sound pressure gives rise to unit particle velocity. The dimensions of the rayl are newton second metre^{-3} in MKS units or dyne second cm^{-3} in CGS. In both instances the unit is equal to the product of the density of the gas and the velocity of sound in the gas; thus air has a specific impedance of 407 MKS rayl or 40·7 CGS rayl. The unit, at the suggestion of Beranek[9] in 1954, was named the rayl

after the distinguished physicist and Nobel prize winner the third Lord Rayleigh (1842–1919).

Rayleigh (R)

The luminous intensity of the aurora and the night sky may be measured in rayleighs, where one rayleigh is equivalent to 10^6 quanta per square centimetre. The unit, proposed[10] in 1956, is named after the fourth Lord Rayleigh (1875–1947) who included in his numerous scientific achievements a thorough investigation of glow discharges in gases. The luminous intensity of the night sky is about 250 R, whereas that from an auroral display lies between 1 and 1000 kR.

Rayleigh number (Ra)

A dimensionless number given by the expression $l^3 g\gamma\Delta\theta\rho/\eta k$ where l denotes length, g is the acceleration due to gravity, γ is the cubic expansion coefficient, ρ is the density, η is the dynamic viscosity coefficient, k is thermal conductivity and $\Delta\theta$ is the temperature difference.

Reactivity units

Reactivity is a number which is defined as K_{ex}/K_{eff}, where K_{ex} is the excess reproduction constant and K_{eff} is the effective reproduction constant[11]. The reactivity of a nuclear reactor is of the order of 10^{-5}; this is called a ppc (pour cent milli) in France and a millik in Canada. In Great Britain a reactivity of 10^{-2} is called a nile.

In the USA there are two units, an inhour and a dollar. The former was proposed by Fermi[12] in 1947 and was derived from the expression *inverse hour*, since it may be defined as the reciprocal of the reactor time in hours. The dollar is the reactivity at which a chain reaction is self-sustaining.

Réaumur scale

See **Temperature scales**.

Redwood second

A unit used in viscometry. (*See* **Second (viscometry)**.)

Refractive index

If a plane electromagnetic wave of wavelength λ falls on a plane boundary between two homogeneous media 1 and 2, the sine of the angle between the normal to the incident wave and the normal to the surface bears a constant ratio to the sine of the angle between the normal to the refracted wave and the surface normal. This constant is known as the refractive index for refraction from medium 1 to medium 2. It is also equal to the ratio of the velocity of light in medium 1 to that in medium 2. When medium 1 is a vacuum the constant is often called the absolute refractive index of the second medium. This constant n is related to the permittivity ε and permeability μ by the relationship $n^2 = \varepsilon\mu$.

The constancy of the ratio of the sines was discovered in 1613 by W. R. Snell (1580–1626) and is known as Snell's Law.

Relative biological effectiveness factor

This is the ratio of the absorbed dose of a standard ionizing radiation (usually hard filtered 200 k V X-rays) that produces a particular biological effect to the absorbed dose of the radiation under consideration that produces the same effect under otherwise identical conditions. In 1968 the International Commission on Radiological units recommended its use only in radiobiology.

Rem

See **Roentgen equivalent man**.

Rep

See **Roentgen equivalent physical**.

Reyn

The unit of dynamic viscosity in the English or foot pound second system. It is named after Osborne Reynolds (1842–1912) and is used in lubrication. The name was suggested in 1961[13].

Reynolds' number (Re)

Reynolds number is one of the more important of the non-dimensional parameters of fluid motion. In 1883 Osborne Reynolds (1842–1912) showed the flow of a fluid in a tube could be described by a dimensionless constant[14]. This constant, which is now called Reynolds' number is given by vl/v, where l has the dimensions of length, v is the fluid velocity and v is the kinematic viscosity. Subsequent work has extended Reynolds' ideas to all cases of fluids in motion. In general low values of Re show that the viscous forces are predominant in controlling the flow whereas higher values indicate that the inertial forces are more important.

When the fluid is moving through a curved pipe its flow continues to be governed by Reynolds' number, but in this case its numerical value is changed by taking into account the ra i of cross section (R) and of curvature (r) of the tube. This derived constant is sometimes called the Dean number and is equal to $\frac{1}{2}Re(R/r)^{1/2}$.

The movement of a fluid between two rotating cylinders of radii R_1 and R_2 may be described by a dimensionless constant based on Reynolds number and known as the Taylor number (Ta), where $Ta = 2Re^2(R_1 - R_2)/(R_1 + R_2)$.

Rhe
The unit of fluidity is the rhe, the name of which is derived from the Greek word rho, to flow. Fluidity (θ) is the reciprocal of dynamic viscosity and the rhe is the reciprocal of the dynamical unit. Bingham, who first used the unit in 1928[15], took the rhe to be the reciprocal of the centipoise. This definition is also used by the British Standards Institution but in *Van Nostrand's Scientific Encyclopedia* (1958) it is defined as the reciprocal of the centistokes.

Rhm
The rhm or roentgen-per-hour-at-one-metre is the unit of effective strength of a gamma ray source such that, at a distance of 1 metre in air, the gamma rays produce a dose rate of 1 roentgen per hour. The unit, with its name, was proposed[16] in 1946 for the quantitative comparison of radioactive sources for which disintegration rates cannot easily be determined. The rhm is of the same order of magnitude as the curie (q.v.).

Richardson number
See **Froude number**.

Richtstrahlwert
An electron-optics unit of brightness. It equals the current density per unit solid angle.

Roc, rom
Names proposed in 1964 for the CGS and MKS units of electrical conductivity[17]. The names are derived from the initial letters of *reciprocal ohm centimetre* and *reciprocal ohm metre* respectively. If σ_{roc} and σ_{rom} denote electrical conductivities $\sigma_{roc} = 100\sigma_{rom}$.

Rockwell number
A hardness number (q.v.).

Rod
A unit of length equal to $16\frac{1}{2}$ feet. The unit is also called a perch or pole. The name rod dates from about 1450.

Roentgen (r)
The roentgen is the unit of radiation exposure. It is defined as the quantity of X-ray or γ radiation such that in the air the associated corpuscular emission per $1\cdot293 \times 10^{-6}$ kg of air produces a quantity of electricity equal to one electrostatic unit. The value $1\cdot293$ kg m^{-3} is the density of dry air at s.t.p. This definition is often converted to read $2\cdot58 = 10^{-4}$ coulomb per kg of air.

The idea of using the ionization produced by 1 cubic centimetre of air as a measure of the output of an X-ray tube was suggested by Villard[18] in 1908. The roentgen, which is called after W. C. Roentgen (1845–1923), the discoverer of X-rays, was proposed at the 1928 Radiology Conference in Stockholm[19]. In the 1928 definition the quantity of air was described in terms of volume. In the present definition, adopted nine years later[20], the quantity of air is defined in terms of mass. In the years between 1928 and 1952 the roentgen used at the National Physical Laboratory in England was about 8 per cent smaller in magnitude than the one used in medicine, but in 1952 the unit used by the NPL was altered so that the medical and the physical roentgens have the same value. At one time the roentgen was used in clinical work to describe both exposure and absorbed dose, but in 1956 it was decided to use the rad for dose and the roentgen for the unit of radiation exposure[21]. Relating to air an exposure dose of a roentgen corresponds to 8.38×10^{-3} J absorbed per kg, 1.610×10^{15} ion pairs produced per kg and 6.77×10^{10} MeV absorbed per m^3.

Roentgen equivalent man (mammal) (rem)

This unit is the quantity of ionizing radiation such that the energy imparted to a biological system per gram of living matter by the ionizing particles present in the locus of interest has the same biological effectiveness as one rad of 200 to 250 kilovolt X-rays[22]. The rem is sometimes called the dose equivalent and is the general unit used in radiological protection (*see* **Rad**); it takes into account the different types of ionizing radiation. 1 rem = 10^{-2} J Kg^{-1} = 0·01 sievert. The unit was proposed and named by H. M. Parker[23] about 1950.

Roentgen equivalent physical (rep)

The dose of radiation which, when delivered to a given mass of tissue, produces the same energy conversion (about 8.38×10^{-6} J) as one roentgen delivered to the same mass of air. The unit was at one time called a tissue roentgen and it has also been called a Parker after H. M. Parker (1910–), who proposed it about 1950[23].

Rossby number

This gives the ratio of the inertial to the Coriolis forces in a moving fluid.

Rowland

A unit of wavelength equal to about 0·1 nm. It was adopted[24] by H. A. Rowland (1848–1901) because of a small error in the wave-length of the lines on Ångström's map of the solar spectrum; this error was due to Ångström's assumption that the Uppsala standard of length was 999·94 mm whereas it was 999·81 mm. The rowland unit was used extensively between 1887 and 1907; it is now obsolete, having been replaced first by the ångström[25], and then by the nanometre.

Rum
A name suggested facetiously in 1934 for the bar[26] which was the unit of pressure equal to 10^{-1} Nm^{-2} when it was proposed to use the bar equal to 10^5 Nm^{-2} bar.

Rutherford
A unit of radioactivity which is defined as the quantity of radioactive material which undergoes 10^6 disintegrations per second. 37 rutherfords = 1 millicurie.

The unit was named after Lord Rutherford (1871–1937), the founder of nuclear physics, by Curtis and Condon[16] in 1946 and was approved by the American National Research Council in 1949[27].

Rydberg
A name suggested by Candler[28] in 1951 for the unit of wave number. It is equal to the number of wavelengths in a centimetre and is named after J. R. Rydberg (1854–1919) who was well known for his work on atomic spectra. The unit is now called the kayser (q.v.).

Rydberg constant
See Appendix 1.

SAE numbers

These provide an arbitrary classification of lubricating oils based on viscosity. They were first introduced by the American Society of Automotive Engineers in 1913. Table 9 gives the relationship between SAE numbers and viscosity.

TABLE 9

SAE No.	Viscosity range Saybold Universal Seconds			
	At 0° F		At 210° F	
	min	max	min	max
5W	—	4000	—	—
10W	6000	12000	—	—
20W	12000	48000	—	—
20	—	—	45	58
30	—	—	58	70
40	—	—	70	85
50	—	—	85	110

Sabine

The unit of sound absorption in architectural acoustics. The unit which is equal to the absorption due to 1 square foot of a perfectly absorbing surface was introduced in 1911 by Wallace C. Sabine (1868–1919)[1], the pioneer of architectural acoustics who used an open window as a perfect absorber. The name sabin was given to the unit in 1937 by the American Acoustical Society[2] but previous to this it had been called the total absorption unit (Sabin) by Stanton, Schmid and Brown[3] in 1934. It is recognized by the American Standards Association.

Absorption, volume and reverberation time are connected by the relationship $a = 0.161\ v/T$, where a is the absorption in sabins, v is the volume (feet ^{3}) and T is the time (seconds) required for the intensity of the sound to decrease by 60 dB, i.e. to become inaudible. There is, as yet, no name for the metric unit corresponding to the sabin.

Saffir-Simpson Damage Potential Scale

A scale used by weather services in which numbers from 1 to 5 separate the potential for wind and storm-surge damage from a hurricane in progress[4]. In the United States the numbers are made available to public safety officials when a hurricane is within 72 hours of landfall. It resulted from the studies of J. H. Saffir an American consulting engineer and R. H. Simpson a former director of the National Hurricane Centre, Miami, Florida. It was first introduced in 1975[5].

The scale is given in Table 10.

TABLE 10   The Saffir-Simpson Damage Potential Scale

Number (category)	Central pressure millibars	Wind mph	Surge ft.	Damage
1	≥980	74–95	4–5	minimal
2	965–979	96–110	6–8	moderate
3	945–964	111–130	9–12	extensive
4	920–944	131–155	13–18	extreme
5	<920	>155	>18	catastrophic

In the United States two hurricanes out of 126 have reached 5 in the period 1900–1974. They affected Florida in September 1935 and Louisana and Mississippi in 1969.

Saunder's theatre cushion

The earliest unit of absorption. It was the standard used by Wallace C. Sabine[4] in his pioneer work on the acoustics of buildings in 1896. The original 'unit' was a cushion from the Saunders theatre at Harvard and this cushion is now preserved by the Acoustical Society of America.

Savart

A unit of musical interval. (*See* **Musical scales.**)

Saybolt second

A unit used in viscometry. (*See* **Second (viscometry).**)

Schmidt number (Sc)

A dimensionless transport number[6] given by $\eta/\rho D$ where η is the dynamic viscosity coefficient, ρ is density and D is the diffusion coefficient.

Screw threads

Screw threads are classified by their diameter, the number of threads N per inch or per centimetre and their form. Sometimes the pitch is given, which is the reciprocal of N. The form is the shape of one complete profile of the thread between corresponding points at the bottom of adjacent grooves as shown in axial plane section. A thread is right-handed if, when assembled with a stationary mating thread, it recedes from the observer when rotated in a clockwise direction. A left-handed thread recedes when rotated anti-clockwise.

Attempts have been made at various times since 1841 to get an agreed form for screw threads. Nevertheless there are over 60 different forms, about 10 of which are frequently used.

In the United Kingdom[7] the commonly used threads are the Unified, the Whitworth, the British Association and the British Standard Cycle.

The Unified series covers screws of 1/4 inch or more in diameter, with N varying from 36 to 4. There is a coarse and a fine series. A screw designated 1/4–20 UNC indicates a Unified coarse screw thread, diameter 1/4 inch, 20 threads to the inch. The Unified series came into use during the 1939–45 war.

The Whitworth series was proposed by Sir Joseph Whitworth[8] in 1841. The British Standard (BS) Whitworth covers the diameter range 1/8 inch upwards, with N varying from 40 to 2·5. The BS Whitworth fine and BS Whitworth pipe are also in the Whitworth series. The former starts with a diameter of 3/16 inch and proceeds upwards in steps of 1/32 inch with a range of N from 32 to 4. The pipe series covers the diameter range 1/8 to 4 inches but N is limited to the values 28, 19, 14 and 11. The pipe series dates from 1905.

The British Association System[9] originated in 1884 due to the necessity of having a standard system for the small screws required in scientific apparatus. In this system the diameters of the screws are in the range 0·236 to 0·01 inches and N varies from 25 to 350. The screws are classified 0 to 25, the usual series ranging from 0 to 10. An 0 BA screw has 25·4 threads to the inch and a diameter of about 0·236 inch. The corresponding figures for 10 BA are 72·6 and 0·067 inch.

The Cycle thread covers the ranges 1/8 inch to 3/4 inch in steps of 1/32 inch or multiples of 1/32 inch and N is limited to the values 40, 32 and 26. A screw designated 1/4–26 BSC indicates a diameter of 1/4 inch, 26 threads per inch, British Standard Cycle.

In the USA the oldest system is the American Standard screw thread which was introduced by Sellers in 1864. This is sometimes called the Franklin system because its use was first sponsored by the Franklin Institute. Other American screw thread systems include the Society of Automotive Engineers (SAE), the American Standard Coarse and the American Standard Fine.

There are three widely used metric threads. The oldest is the Delisle or

Lowenhertz which was developed in Germany in 1873. The others are the Thury and the Système International (SI), the latter having been confirmed at a conference held at Zurich in 1898. The SI system as used today differs a little in detail from what was originally agreed. For diameters in the range 6 to 80 mm, the related pitches are identical with those agreed at the 1898 conference except that the pitches for the 72, 76 and 80 mm threads are 6·5, 6·5 and 7 mm respectively instead of 6 mm. For diameters above 80 mm the French (CNM 3), the German (DIN 14) and the Swiss (VSM 12004) standards differ from the Zurich standards but they all have a uniform pitch of 6 mm. For screw diameters of less than 6 mm, some of the metric countries use their own national gauges; thus the French (CNM 132·133) differs from the German (DIN 13), but the latter is the same as that used in Switzerland (VSM 12002).

Other types of thread include the square and the acme; both of these are used mainly in worm wheels for power transmission.

In 1964 the British Standards Institution[10] decided to use only ISO metric, and ISO unified inch, threads. This decision should make all other screw threads obsolete in the United Kingdom.

Secohm
A name proposed by Ayrton and Perry for the practical unit of inductance in 1887[11]. The secohm was equal to the product of 1 legal ohm and 1 second and its magnitude was about the same as a henry (q.v.).

Second (angle)
The sixtieth part of a minute of arc.

Second (s) (time)
This is the fundamental unit of time in all systems of units. It is defined as the duration of 9 192 631 770 periods of the radiation corresponding to the transition between two hyperfine levels of the ground state of the caesium 133 atom. This definition was adopted by the 13th CGPM (1968)[12]. Previous to this the unit of time had been identical to the astronomical second of ephemeris time which was defined as 1/315 569 259 747 of the tropical year 1900. This second is now known as the ephemeral second and the one defined with reference to the caesium atom as the SI second.

Second (viscometry) Redwood, Saybolt and Engler
In certain industrial capillary viscometers the kinematic viscosities of fluids are found by noting the time in seconds for a given quantity of the fluids concerned to pass through a capillary tube[13]. The instruments used are Redwood viscometers No. 1 and No. 2 (Great Britain), the Saybolt Universal and the Saybolt Furol (*fu*el and *ro*ad oil) viscometers (USA) and the Engler viscometer (continental Europe). If the absolute values of the kinematic

viscosities be required, these can be obtained from tables which correlate the flow times with the viscosity expressed in centistokes. The flow times are given in Redwood seconds, Saybolt seconds and Engler seconds. The first two are used with Redwood and Saybolt viscometers. The Engler second is the time for 200×10^{-6} m^3 of oil to flow through an Engler viscometer.

The order of magnitude of the times of flow are given by the following examples. A coefficient of viscosity of 10 cSt is equivalent to 51·7 seconds with a Redwood No 1 viscometer at 75°F, 58·8 seconds with a Saybolt Universal at 100°F, and about 50 Engler seconds. The Redwood No. 2 and the Saybolt Furol viscometers are used for liquids having a viscosity coefficient greater than 100 cSt. For these 100 cSt corresponds to 40 seconds with a Redwood No. 2 and 48·5 seconds with a Saybolt Furol at 122°F. Conversion tables for Redwood viscometers are given in *Standard Methods for Testing Petroleum and its Products* (7th Edn, 1946) and for Saybolt viscometers in *The Book of Standards*, Part 3 (1944) of the American Society for Testing Materials.

Sensation unit

A unit of loudness suggested[14] in 1925. If P_0 be the sound pressure level which can just be detected by the ear, then P the sound pressure under examination is S sensation units greater than P_0, where S is derived from $S = 20 \log_{10}(P/P_0)$. The unit was based on a one-for-one relationship between sound pressure and loudness, the former being objective and the latter subjective. This is a false assumption and the unit has been wisely forgotten.

Shackle

A unit used for measuring the length of cable or studded chain. The name dates from the sixteenth century and originally referred to the hook or ring to which the cable was attached. In time, it was applied to the distance between two such shackles, thus the reference in a nineteenth century seamanship manual 'length of the bower cable is usually 12 shackles, each shackle being 12·5 fathoms'. In 1949 the British Navy replaced the 12·5-fathom shackle by one of 15 fathoms for describing the length of cables.

Shed

A proposed unit for nuclear cross-section equivalent to 10^{-52} m^2 per nucleus or 10^{-24} barn. The unit is so very small that to date it appears to have been little used. It is mentioned in *The International Dictionary of Physics and Electronics* (Macmillan 1956).

Sherwood number (*Sh*)

A dimensionless number used in heat transfer[15]. Its value is given by $Sh = j_m(Re)(Sc)^{1/3}$, where j_m is the mass transfer factor, *Re* Reynolds' number (q.v.) and *Sc* is Schmidt's number (q.v.).

Shore scleroscope number
A unit of hardness. (*See* **Hardness number (solid).**)

SI unit of exposure to radiation
This is expressed in coulomb kilogram^{-1}; 1 SI unit = 3876 roentgens. No name has been suggested for the unit, which is already becoming obsolete.

Shower unit (s)
This is a unit of length used in cosmic ray work[16]. It is the mean path length required to reduce the energy of a charged particle by 50 per cent. The shower unit s is related to X_0 the radiation length or cascade unit by $s = X_0 \ln 2$. The value of s in air is 230 metres, in water it is 300 mm and 35 mm in lead. The radiation length, sometimes called a radiation unit, dates from about 1940; the terms cascade and the shower units were first used about a decade later.

Siegbahn unit (X unit)
The siegbahn, or X unit, is used for describing wavelength in X-ray spectroscopy. It equals $100 \cdot 202 \pm 0 \cdot 004 \times 10^{-15}$ m or approximately 10^{-4} nm. The unit is defined in terms of the spacing of the (200) planes in a calcite crystal which at 18°C is taken to be 3029·45 Siegbahn units apart. It was introduced in 1925[17].

Siegbahn chose his unit to be of the order of 10^{-4} nanometre but insisted it should have a distinctive name so that when measuring technique improved the discrepancy beween his unit and 10^{-4} nm could be clearly indicated by a numerical factor. This factor has been given the symbol Λ and its current value[18] is 1·00208.

Siemens (S)
The siemens is the SI unit of electrical conductance and it has the dimensions of ohm^{-1}. The unit, although approved by the IEC in 1933[19], was not adopted until the 14th CGPM (1972). It is named after E. W. Siemens (1816–92) who did much to develop the electric telegraph and pioneered the use of a column of mercury as a standard of resistance in 1860. The unit was originally called the mho (q.v.).

Sieve numbers
Sieving provides a rapid and convenient method for the grading of particles according to size. In recent years the mesh in sieves has been standardized. In Great Britain the scale used (BS 410: 1943) gives the number of meshes to the linear inch; thus a mesh number of 200 indicates 200 meshes to the inch made from wire 2×10^{-3} inch in diameter with a distance of 3×10^{-3} inch between each wire. A similar scale is used in the United States of America (ASTM, E11–61); here the number, called the sieve number, indicates the size of the aperture. The aperture indicated by any particular sieve number has

approximately the same size as that given by the British mesh number; small discrepancies between the two scales occur for the British and American wire gauges do not always coincide. Sieves may also be graded in Tyler mesh numbers or in the Institution of Minerals and Metals (IMM) scale. The former are very similar to the American sieve numbers and the latter is an arbitrary scale the numbers of which increase in value as the sizes of the apertures get smaller. The close correlation between mesh numbers, sieve numbers and the Tyler scale is indicated in Table 11.

TABLE 11

Mesh No.	Aperture inch	μm	Sieve No.	Aperture μm	Tyler No.	Aperture μm	IMM scale
60	$9 \cdot 9 \times 10^{-3}$	251	60	250	60	250	50
100	$6 \cdot 0 \times 10^{-3}$	152	100	149	100	149	80

Sievert (Sv)
The 16th CGPM (1980) adopted the sievert as the SI unit of dose equivalent[20], one sievert representing a radioactive dose equivalent of one joule per kilogram, i.e. 1 Sv = 100 rem. The name adopted for the unit was the same as that suggested by the International Commission on Radiological Protection in 1977; previous to this the unit of dose equivalent had been called the intensity millicurie[21] and was equal to 8·4 r.

Sign
An angular measure equal to 30°. It is derived from the signs of the Zodiac, each of which was equal to a twelfth of a circle – 30°.

Sikes
A specific gravity unit. (*See* **Degree (hydrometry)**.)

Siriometer
An obsolete unit of length first suggested by C. L. W. Charlier (1862–1934); (1 siriometer = 10^{6} astronomical units).

Siriusweit
An obsolete unit of length used by the German astronomer M. H. Seeliger (1849–1924); (1 siriusweit = 5 parsecs).

Skewes number
Kasner and Newman in their book *Mathematics and the Imagination* state that Skewes number[22] gives an indication of the number of primes which they claim to be 10^{3800}. They quote G. H. Hardy's *Pure Mathematics* as their source of information. Skewes number is not mentioned by name by Hardy but he claims that the number of primes is given by 10^{500}, a figure which is

also equal to the total number of games of chess which may be played. The discrepancy in value between the Kasner, Newman and Hardy *Skewes number* exceeds by many powers of ten the largest error reputed to have been made in a lecture. This took place when an astronomer found his volume was incorrect by a factor of 10^{55} – he had worked in cubic parsecs instead of cubic centimetres!

Skot
A suggested unit for poor luminance of a surface equivalent to 10^{-3} apostilb or 10^{-3} lumen per square metre. It was introduced in Germany for measuring 'black-out' lighting during World War II[23].

Slug
A unit of mass in the foot pound system. It is a mass the value of which is numerically equal to the acceleration of standard gravity (q.v.), so that it will assume an acceleration of 1 foot per second per second when subjected to a force of 1 pound weight. The value is 32·1740 pounds. The unit is reputed to have been invented by John Perry[24] in 1890 but the name slug was not used until 1902[25]. The name is no doubt associated with slow progress of the slug. The slug is also called a gee pound.

Snellen
A unit expressing the visual power of the eye. It is named after the Dutch ophthalmologist Herman Snellen (1834–1908) who designed and published optotypes. The unit was suggested in a paper on visual power and visibility in 1951[26].

Sol
A Martian day, which equals 24 hours 36 minutes.

Sone
A subjective loudness unit designed to give scale numbers proportional to the loudness heard by the listener. By definition a simple tone of frequency 1000 Hz and 40 dB above the listener's threshold of hearing produces a loudness of 1 sone; thus the loudness of any source deemed by the listener to be n times the strength of this standard has a loudness of n sones. The relationship[27] between the loudness in sones S of any sound and its loudness level in phons P is given by $S = 2^{(P-40)/10}$, hence 1 sone is equal to 40 phons. The unit was proposed by Stevens and Davis[28] in 1938 and is recognized by the American Standards Association[4].

Soxhlet scale
A scale used in lactometry. (*See* **Degree (hydrometry)**.)

Span
An Anglo-Saxon unit of length. One span $= 0.75$ foot $= 2.286 \times 10^{-1}$ m.

Spat (S)

An astronomical unit of distance equal to 10^{12} m suggested by Callou[29] in 1944. It is claimed the unit will remove the necessity of using large numbers for describing distances – the sun is 0·15 spat from the earth.

Speech interference level scale (SIL)

An arbitrary noise scale used in acoustics. (*See* **Subjective sound ratings**.)

Speed of light (*c*)

At the 17th CGPM (1984) the metre was redefined as the length of the path travelled by light in a vacuum during a time interval of $1/(299\,792\,458)$ second. One of the consequences of this definition is that *c*, the speed of light, has become the second physical constant to be fixed by convention, the other is μ_0, the permeability of free space ($4\pi \times 10^{-7}$ H m^{-1}). Furthermore, since $c^2 = 1/(\mu_0\varepsilon_0)$ the permittivity of free space, ε_0, has also become fixed by convention, its value being $8·854\,187\,818 \times 10^{-11}$ F m^{-1}.

Square degree ($\Box°$) or $(°)^2$

A unit of solid angle equal to $(\pi/180)^2$ steradian.

Square grade $(g)^2$

A unit of solid angle equal to $(\pi/200)^2$.

Stab

A name sometimes used for the metre (q.v.).

Stadia

A unit of distance used in ancient Egypt. In 200 B.C. Eratosthenes made the first reliable measurement of the circumference of the earth by observing that the sun was directly overhead at Syene at noon at the summer solstice and was about 7° from the vertical at Alexandria, a distance of 5000 stadia (925 km) to the north; from this he calculated the circumference of the earth was about 26 000 miles.

Standard atmosphere

The International Civil Aviation Organization (ICAO) standard is equivalent to a pressure of $1·013250 \times 10^5$ Nm^{-2}, 760 torr or 1013·250 millibars. The definition assumes the air to be a perfect gas at 15°C (288·16 K) and at mean sea level. It should be noted that the pressure is defined in terms of newton m^{-2} instead of millimetres of mercury. One standard atmosphere corresponds to a barometric pressure of 29·9213 inches or 760 mm of mercury of density $13·595 \times 10^3$ kg m^{-3}, where the acceleration due to gravity is 9·80655 m s^{-2}. The equivalent value in pounds per square inch is 14·691. The ICAO standard atmosphere was introduced about 1940[30].

Other standard atmospheres which have been used are those of the

National Advisory Committee on Aeronautics (NACA) and the International Commission for Air Navigation (ICAN). The former dates from 1922 and the latter from 1924. In both, pressure is given in terms of a column of mercury 760 mm high at a specified temperature. The main difference between these two standard atmospheres is that in the NACA atmosphere the acceleration due to gravity is taken to be 9.8066 m s^{-2} whereas it is 9.8062 m s^{-2} in the ICAN atmosphere.

Standard cable

The standard cable was a unit of attenuation adopted in 1905 by the National Telegraph Company in the USA and by the Post Office in Great Britain[31]. The unit compared the attenuation produced in the circuit under test with that in a standard cable which was defined as a theoretical cable 1 mile in length, resistance 88 ohms, capacitance 0.054 microfarad, inductance 1 millihenry and leakance 500×10^3 mho. The standard cable produced an attenuation of about 20 per cent for a 800 Hz input. The unit was replaced by the transmission unit (TU) in 1922 and this name was superseded by the bel in 1923[32].

Standard gravity

The acceleration due to gravity varies according to the locality but, for convenience in calculation, the value at Potsdam ($g = 9.81274$ m s^{-1}) has been used as a standard for over 60 years and, from this, local values of g could be obtained by comparing the periodic times of a pendulum swung at Potsdam and at the location concerned. However, in the early 1970s, the 61st CIPM (1972) recommended that, whenever precision measurements involving gravitational forces had to be made[12], the relevant local value of g should be obtained from the International Gravity Standardization Network (IGNS–1971) which had been drawn up by the International Unit of Geodesy and Geophysics at its Moscow meeting in 1971.

Standard illuminants

The well-known phenomenon that colours appear to be different when seen by artificial light has made it imperative that the light used when colours are matched must be specified. In 1931 the CIE[33] defined three standard illuminants which are designated A, B and C. A is the light from a filament at a colour temperature of 2848 K, B represents noon sunlight and C normal daylight. B and C are defined with respect to A, the necessary changes in spectral intensity being produced by filters which are rigorously specified. These standards were originally suggested by the Colorimetry Committee of the Optical Society of America[34] in 1920–21.

Standard volume

Standard volume is the volume occupied by 1 kilogram molecule of a gas at

0°C and at a pressure of 1 standard atmosphere. Its present value is approximately 22·414 m³.

Stanton number (St)

The Stanton number is used in forced convection studies[35]. It is equal to the Nusselt number divided by the product of the Prandtl and Reynolds numbers, i.e. $h/\rho v c$ where h is the coefficient of heat transfer, ρ is the density, v is velocity and c is specific heat. It is named after Sir Thomas Stanton (1865–1931), an authority on gaseous flow.

The Stanton number (St*) used in mass transfer problems is given by $m/tA\rho v$ where m is the mass transferred across area A in time t.

Star magnitude

Hipparchus, the Greek astronomer who lived in the second century B.C., called the twenty brightest stars 'stars of the first magnitude'. The less bright stars were considered to be of higher, i.e. second, third, etc., magnitudes. This system of star classification is in use today. Stars of the first six magnitudes are visible to the naked eye on a clear night, but a telescope is required for seeing stars of greater magnitude.

Star magnitudes are defined by the relationship $\log_{10} E = 0\cdot4\,(1-m)$, where E is the intensity of a star relative to a first magnitude star and m is the magnitude[36]. Thus a star is one magnitude above another when the intensities of the light from them are in the ratio of $1:100^{1/5}$ or $1:2\cdot512$. The illumination produced by a first magnitude star on the earth's surface at the zenith is of the order of 0·9 micro-lux or 2·3 mile-candles[37].

Stat

A rarely used unit of radioactivity equal to $3\cdot63 \times 10^{-27}$ Ci.

Statampere (statcoulomb, statfarad, statohm, statvolt)

The prefix stat denotes CGS electrostatic (e.s.u.) system of units, e.g. statampere is the unit of current in the e.s.u. system. (*See* **CGS units**.)

Stathm

One of the names suggested by Polvani[38] in 1951 for the gram. It is derived from the Greek to weigh.

Staudinger value

A number which is sometimes used to give a value to the molecular weight of a polymer. It is sometimes known as the Staudinger molecular weight but does not represent the actual molecular weight of the polymer, which is usually considerably higher. The number is still used in sales literature and charts are available which relate the number to the real molecular weight. The name arises from the work of H. Staudinger (1881–1965), a Nobel prizewinner for chemistry.

Stefan–Boltzmann constant
See Appendix 1.

Steradian (sr, Ω_0)
A unit of solid (three-dimensional) angular measure. 1 steradian is equal to the angle subtended at the centre of a sphere by an area of surface equal to the square of the radius. The name for the unit seems to have come into use about 1880[39] and was comparatively common by the turn of the century. The surface of a sphere subtends an angle of 4π steradians at its centre.

Stere (st)
A unit of volume equal to a cubic metre. It was extensively used in France for measuring bundles of firewood and was originally promulgated in 1798.

Sthéne (sn)
The fundamental unit of force in the metre, tonne, second system of units. It is the force required to accelerate 1 metric tonne (1000 kg) by 1 metre per second per second; it is equal to 10^3 newtons. The unit was originally proposed by the British Association[40] in 1876 when the name suggested was the funal, meaning a rope. It was subsequently called a sthéne, a name which received legal sanction in France[41] in 1919.

Stigma (σ)
A unit of length equal to 10^{-12} m which was proposed by Callou[29] in 1944 to give reasonable numbers for atomic measurements; thus 1 bohr radius would be 52·9 σ. The unit takes its name from the Greek word for a dot. The same distance has also been called a bicron (q.v.).

Stilb (sb)
This is the unit of luminance (formerly called brightness) or luminous intensity emitted per unit area in a given direction. It is a CGS unit and is equivalent to 1 candela per square centimetre. The luminance of a carbon arc crater is of the order of 16×10^3 stilbs and that of clear sky 0·2 to 0·6 stilb. The name is reputed to have been made up[42] by the French physicist A. Blondel (1863–1938) in 1921.

Stokes (St)
The stokes is the CGS unit of kinematic viscosity. It is defined as $St = \eta/\rho$ where η is the coefficient of dynamic viscosity and ρ is the density of the fluid. Kinematic viscosity determines the type of flow because turbulence occurs at faster rates for fluids with a large kinematic viscosity than for those for which the viscosity is small. In 1928[43] the unit was named after Sir George Gabriel Stokes (1819–1903) at one time Lucasian Professor of Mathematics at Cambridge. Another name used for this unit was lentor[44] (q.v.).

The British Standards Institution recommend[45] that, for the calibration of viscometers, the kinematic viscosity of water should be taken as 1·0038 cSt at 20°C.

Strehl intensity ratio
A criterion of image quality used in lens design: it quantifies aberration. It is defined as the ratio of the luminous intensity in the central maximum of the actual image (i.e. the airy disk for a circular source) to that of an aberration-free image. In Germany it is called the definitionshelligkeit.

Strich
A name sometimes used for 1 millimetre.

Strontium unit
This unit, sometimes called a sunshine unit, gives the strontium 90 contamination of food[53]. It is the number of microcuries of Sr^{90} absorbed in one kilogram of calcium.

Strouhal number (Sr)
A dimensionless number involving the frequency of the vibrations produced in a taut wire by the passage of a current of fluid. The vibrations are called aeolian tones. The number is given either as nd/v or v/nd, where v is the fluid velocity, n the frequency of the note and d the diameter of the wire. It has a value between 0·185 and 0·2 or 5·4 and 5 according to the definition used. Occasionally n is replaced by $2\pi n$.

The relationship was derived by V. Strouhal (1850–1922) in 1878[46] but the phenomenon has been known since the days of the 'aeolian harp'. This instrument was first described systematically in an early seventeenth-century book by Athanasius Kircher. Rayleigh gives the Strouhal number as being equal to $0·195 (1 - 20·1/Re)$ where Re is Reynolds' number, so that if, as in practice, Re is greater than 500, the Strouhal number is 0·195.

The Strouhal number is sometimes called the 'reduced frequency' and is used extensively in work on fast moving fluids as for example the air in the vicinity of the tip of an aircraft propeller[47].

Sturgeon
A name suggested by Sir Oliver Lodge for a practical unit of magnetic reluctance in 1892[48]. It was named after W. Sturgeon (1783–1850), one of the inventors of the galvanometer.

Subjective sound ratings
There are several scales by which the subjective loudness of a sound may be correlated with its physical characteristics of intensity and frequency[49].

Beranek Scale. In this scale, proposed in 1956[50], noises are arranged into six arbitrary categories in a manner somewhat analogous to Mohs' scale of hardness. The categories are: very quiet, quiet, moderately quiet, noisy, very noisy, and intolerably noisy.

SIL Scale. In 1950 a scale was suggested in which noises are arranged according to their Speech Interference Levels[51]. The latter is defined as the average sound pressure in the three octaves between 600 and 4800 Hz. On this scale the very quiet on the Beranek scale would correspond to a SIL of zero, moderately quiet would have a SIL of about 45 dB and very noisy would be about 70 dB.

ELR Scale. In the Equal Listener Response scale Beranek and Miller[52] attempted to get an average response of a listener to a sound when allowance was made for the apparent increase of intensity of a noise as its frequency increases.

Two other scales often used are the loudness level scale (*see* **Sone**) and the perceived noise level (*see* **Noy**).

Sumner unit

This is an enzyme activity unit referred to urease. The products of enzyme-catalysed hydrolysis are mainly ammonium and bicarbonate ions, hence a Sumner unit is the amount of enzyme that will liberate 1 milligram of ammonia–nitrogen at 20°C in 5 minutes. The unit takes its name from J. B. Sumner (1877–1955), the biochemist who was the first to isolate enzymes and who was awarded the Nobel prize for chemistry in 1946. 1 Sumner unit = 14·28 International Union of Biochemistry activity units; see International Biological Standards.

Sunspot number

A number devised by R. Wolf of Zurich in 1852 to describe sunspot activity. The number is sometimes called the Wolf number. It is defined by the relationship

$$R = k(10g + f)$$

where R is the sunspot number, k is a constant depending on the instrument used, g is the number of disturbed regions and f is the total number of sunspots.

Surface colour classification

The luminosity of a surface colour is described by its hue, chroma and value. Hue is the quantity which distinguishes one colour from another and thus blue and green are hues whereas black and white are not. The purity of a colour is called its chroma and determines the place of the colour on the chroma scale. Colours which contain no grey are said to be pure. The shade of grey in a colour gives the value of the colour. Shades are arranged on a scale called the grey scale which goes from one (black) to ten (white).

Colour may be specified by a three-dimensional structure the dimensions of which represent hue, chroma and value respectively[54]. Such systems have been devised by Munsell for artists in 1907, by Ridgeway for ornithologists in 1912 and by Ostwald in 1918.

Surface flatness (or roughness)
In workshop practice there are two systems for expressing surface flatness[55]. These are the M (mean line) and the E (envelope) systems. Flatness in both systems is expressed as the distance of some characteristic of the surface from a suitable datum line. The surface characteristic and datum line are rigorously defined in standards issued by the major countries. The unit used for distance is either the micro-inch or micrometre.

The optical industry[56] expresses flatness as the distance between the highest and lowest parts of a surface and measures this in terms of the wavelength λ of the mercury green line (546 nm). An optical flat is a glass surface worked to within $\lambda/60$ or 9 nm.

Survival ratio
The ratio of virus, bacterial or enzyme activity after irradiation by ionizing radiation to that before irradiation.

Svedberg (S)
The velocity of sedimentation of molecules per unit accelerating field is known as the sedimentation coefficient. It is used in work with analytical centrifuges and is defined by the equation $S = v/\omega^2 x$ where v is the velocity of the boundary between the solution containing the molecules and the solvent, x is the distance of the boundary from the axis of rotation and ω is the angular velocity in radians per second. Values of S lie in the range 0·025 to 50×10^{-12} second. The unit 10^{-13} second is called a svedberg (1884–1971) the Swedish pioneer in the use of the ultracentrifuge. It first appeared in 1942[57].

Sverdrup
The flow of ocean currents is measured in sverdrups, where 1 sverdrup represents a flow of a million cubic metres per second. It is named after H. U. Sverdrup (1888–1957), director of the Norwegian Polar Institute who accompanied Amundsen on several of his expeditions to the Arctic.

T

t number
This relates the photographic lens stop setting to light transmission.

Talbot
An MKS unit of luminous energy first used in 1937[1]. A joule of radiant energy having a luminous efficiency of x lumens per watt has a luminous energy of x talbots. It is named after W. H. Fox Talbot (1800–77) who discovered the principle of the flicker photometer in 1834. (*See* **Light watt**.)

Temperature scales
Celsius. This mercury-in-glass scale was devised as early as 1710 and was used by Linnaes at Uppsala certainly before 1737. The zero of the scale represents the melting point of ice and the boiling point of water is taken to be 100 degrees. In continental Europe the scale has always been known as the Celsius scale in the mistaken belief that it was invented by Anders Celsius (1701–44), whereas Celsius proposed a scale which had zero for the boiling point of water and 100 for the melting point of ice. The scale was inverted by J. P. Christen (1683–1755) in 1743. In England the scale was originally called the Centigrade scale but this name was abandoned[2] in favour of Celsius in 1948.

Fahrenheit. This scale was composed by G. D. Fahrenheit (1686–1736) between 1710 and 1714. Three fixed temperature points were used – the temperature of an ice and salt mixture, the freezing point of water and normal human temperature – which were taken to be 0, 32 and 96 respectively. It is mere coincidence that the temperature interval between the freezing (32°F) and boiling (212°F) points of water is 180° when expressed in the Fahrenheit scale.

Réaumur. An arbitrary scale in which the freezing and boiling points of water are taken to be 0 and 80°R respectively. Réaumur deduced his scale in 1730 from the thermal expansion of an alcohol and water mixture. When he considered the 'length' to be 1000 units at the ice point he found that at the boiling point, the length had expanded to 1080 units, hence the peculiar figure of 80 in his scale.

Thermodynamic. In this scale the temperature is considered to be proportional to the energy contained in a given volume of a perfect gas. The temperature is zero when the energy is zero and the temperature of the triple point of water is defined as equal to 273·16 degrees when these are degrees on the Celsius temperature scale. This is equivalent to 459·69 degrees on the Fahrenheit scale. Thermodynamic temperatures are generally expressed in degrees Kelvin or degrees Rankine. Kelvin temperatures represent the same temperature interval as those on the Celsius scale and are obtained by adding 273·16 to the Celsius temperature; Rankine temperatures have the same temperature interval as those on the Fahrenheit scale and are obtained by adding 459·69 to the Fahrenheit temperatures. The scales are named after Lord Kelvin[3] (1824–1907) and W. J. M. Rankine[4] (1820–72), both of whom held chairs at Glasgow University and who, along with Joule, established the thermodynamic temperature scale in the middle of the nineteenth century. In Glazebrook's *Dictionary of Applied Physics* (1923) there is a suggestion that the present Kelvin and Rankine scales should be called the tercentesimal (273≈300) and quinquentesimal (1460≈500) scales respectively. The Kelvin scale is also called the Absolute scale. In 1968 the CGPM[5] decided to replace the term degree Kelvin (°K) by the single word kelvin, for which the abbreviation is K.

International practical temperature scale (IPTS 68). This scale was introduced in 1927 to overcome the practical difficulties of the direct realization of temperature by gas thermometry and to provide a practical temperature scale which could be easily and accurately reproduced. Initially the scale had six fixed points, but it was revised in 1948 and extended in 1968 when its lowest fixed point was reduced from 54·261 K, (boiling water of oxygen) to 13·81 K, (triple point of hydrogen) but its upper limit remained unchanged at 1064·43°C, the freezing point of gold. Between these two limits there are nine accurately defined fixed points, one of which is the triple point of water (0·01°C)[6]. Temperatures on the 1968 scale (IPTS 68) are written with the subscript 68, thus $t_{68} = T_{68} - 273·15$ K, where t_{68} and T_{68} are in degrees Celsius and Kelvin respectively.

In the late 1970s the IPTS 68 scale was extended provisionally to 0·519 K, the superconducting transition point of cadmium, the new scale being called the EPT 76 scale. Finally, it was announced in 1982 that this scale would probably be extended in the near future to cover temperatures between 0·015 K, the (superconducting transiton point of tungsten), and the melting point of tantalum (3000°C).

A brief account of the IPTS and the Chappius constant volume scale which it replaced was given by J. A. Hall in 1967.

Helium scale. In 1958 international agreement[7] was obtained to use the vapour pressure of helium 4 as an indication of temperature in the region 1 to 5·2 K. This scale is called the helium scale.

Curie temperature scale. This is sometimes used for indicating temperature in the vicinity of absolute zero[8]. It is based on Curie's law which states the susceptibility of a paramagnetic material is approximately proportional to its absolute temperature. Curie temperatures are sometimes called magnetic temperatures[9].

Tempon
A unit of time equal to about 10^{-23} second. (*See* **Chronon.**)

Tenth metre
A name occasionally used for the ångström – 10^{-10} metre.

Tesla (T)
The name used by the SUN Committee for the unit of magnetic flux density in the MKS system of units in 1961. 1 tesla represents a flux density of 1 weber per square metre. The unit is named after N. Tesla (1857–1943) who invented the Tesla coil in 1892. The name was approved by the IEC in 1954.

Tex
A unit used in the textile industry for describing the mass per unit of length of a yarn. (*See* **Yarn counts.**)

Therm
A unit of heat in the British system equal to the heat required to raise 1000 pounds of water through 100°F. The unit was defined in the Gas Act[10] of 1920 as being equal to 10^5 Btu.

The British Association at their Bath Meeting[11] in 1888 decided to use the name therm for the heat required to raise 1 gram of water through 1 degree Centigrade (Celsius) but eight years later the calorie was adopted as the name of the unit and the therm went into a state of suspended animation until it was used for the 10^5 Btu unit of 1920.

In 1895 Joly[12] suggested the therm be used to represent the quantity of heat required to change 1 gram of water at 100°C to steam when the pressure was 1 atmosphere. This would give a value of 1·8 millitherms as the specific heat of water.

Thermal ampere, ohm, farad, henry
The advantage of using electrical analogies in heat has been shown by Bosworth[13]. Thus temperature difference, θ, is considered to be analogous to potential difference and the quantity of heat q (joules) flowing per second is

analogous to current and is expressed in thermal amperes. A thermal ampere is equivalent to an entropy flow of 1 watt per degree Celsius. Thermal resistance is θ/q and thermal capacitance is qt/θ where t is the time in seconds. The unit of thermal resistance is the thermal ohm and a body has this resistance if unit temperature difference causes an entropy flow of 1 watt per degree Celsius. The thermal resistances of copper and iron are $7\cdot56 \times 10^3$ and $46\cdot9 \times 10^3$ thermal ohms m^{-1} respectively. The thermal ohm was proposed by White[14] in 1938.

The unit of thermal capacitance is the thermal farad and a body has a capacitance of 1 thermal farad if an amount of entropy equal to 1 joule per degree Celsius added to the body raises its temperature by 1°C. The capacitances of 1 cubic metre of copper and of water at 20°C are $11\cdot6 \times 10^6$ and $14\cdot3 \times 10^6$ thermal millifarads respectively.

The thermal analogy of inductance is based on the fact that the system of convection currents surrounding a hot body immersed in a fluid contains energy. Such a fluid has a thermal inductance of 1 henry if an entropy flow of 1 watt per degree Celsius is associated with a hydrodynamical energy of 1 joule. So far no numerical data have been published to indicate the magnitude of the thermal henry. The idea of thermal inductance was proposed by Bosworth[15] in 1946.

Thermie (th)
The basic unit of heat in the metre tonne second system. It represents the heat required to raise the temperature of 1 tonne of water by 1 degree Celsius and is therefore equal to $4\cdot185 \times 10^6$ joules or 3967 Btu. It was legalized in France in 1919[16].

Thou
A colloquial term for a thousandth of an inch. It has been in use in Great Britain since Victorian times.

Thousandth mass unit (TMU)
An arbitrary unit of energy derived from Einstein's equation $E = mc^2$, where m is the atomic mass unit multiplied by 10^{-3} and c the velocity of light; thus $1 \text{ TMU} = 1\cdot492 \times 10^{-13}$ J.

Tissue roentgen
A name formerly used for the roentgen equivalent physical (q.v.).

TME
The abbreviated form of the name used in Germany and Switzerland (*T*echnische *M*ass *E*inheit – Engineering Mass Unit) for the mass which is accelerated by 1 metre per second per second by a force of 1 kilogram weight.

Its adoption in the English-speaking world was suggested[17] in 1960. There was a proposal to call it the massau after a Belgian engineer of that name.

Tog
A unit suggested in 1946 for the measurement of the insulation of clothing [18]; it is defined as $1 \text{ tog} = 0.100°C \text{ m}^2\text{W}^{-1} = 0.116°C \text{ m}^2\text{h kcal}^{-1} = 0.645$ clo (q.v.).

Toise
A former unit of length used for geodetic measurement. There are two toises – la Toise du Pérou fabricated prior to 1735 measuring 1·94909 m, 6 Paris feet or 6·395 English feet and the Toise du Chatelet. The original Peruvian toise, which was constructed to allow the assessment of the length of the seconds pendulum at the equator, still exists[19]. It was copied 80 times in 1766 and was used in France until replaced by the metre.

Tolerance unit
This unit expressing the degree of tolerance allowed in engineering for fitting cylinders into cylindrical holes. Sixteen grades of tolerance are possible and these are designated $IT.1$ to $IT.16$. These grades are called fundamental tolerances and are multiples of the fundamental tolerance unit i, where if D is the diameter

$$i(\mu m) = 0.45\ D^{1/3} + 0.001\ D$$
or
$$i(0.001 \text{ in.}) = 0.052\ D^{1/3} + 0.01\ D$$

In these expressions D is given in millimetres and inches respectively.

The unit is approved by the International Organization for Standardization[20].

Ton
A unit of mass in both the metric and the foot pound systems. In the United Kingdom the ton is equal to 2240 avoirdupois pounds; this mass is also called a *long* ton. In the USA the *short* ton of 2000 pounds is used and in the metric system the tonne is 1000 kilograms (2205 pounds). The name of the unit is derived from the same source as that of a tunne of wine, a cask which held about 250 gallons. The ton, as a unit of mass was in use in the late fifteenth century[21] and is mentioned in an act of Henry V (1422) in which the measurement of coal from Newcastle is defined.

Ton of refrigeration
Refrigeration is usually computed in Btu per minute[22]. In large commercial installations, however, the Btu is too small and the plants are rated in tons of refrigeration. This unit is of American origin and represents

the heat required to freeze a short ton (2000 lb) of water at 0°C in 24 hours, i.e. it represents 288 000 Btu of refrigeration in 24 hours or 200 Btu per minute. The unit is recognized by the American Society of Mechanical Engineers In British Standards there is a definition for a ton of refrigeration in which the long ton (2240 lb) is employed, so that the unit represents 322 560 Btu per day, but it is little used, because British refrigeration engineers prefer to work in kilocalories per second. One kilocalorie per second is equal to 342 860 Btu in 24 hours. In continental Europe the unit of refrigeration was, until recently, known as the frigorie (q.v.).

Ton (shipping)

The size of a vessel may be described in terms of either displacement, gross or net tonnage. Displacement tonnage is the number of tons of sea water displaced by a vessel when charged to its load line, i.e. it is the mass of the vessel and its contents expressed in tons. Displacement tonnage is generally used for describing the size of warships. Gross registered tonnage is the capacity, in cubic feet, of the space enclosed by the hull and deck houses of a vessel divided by 100. It is used mainly for merchant vessels. Net registered tonnage is the gross tonnage less deductions for the spaces occupied in working the vessel, e.g. engine room, bunker space and crew accommodation. Canal and harbour dues are generally based on net tonnage, which gives an indication of the revenue-earning potential of the vessel. Yachts, when classified for racing, are described by Thames measurement. This scale was devised by the Royal Thames Yacht Club in the 19th century and is based on the formula

$$\frac{(\text{length} - \text{beam}) \times (\text{beam})^2}{188}$$

where the length and beam are measured in feet.

Tonne

A unit of mass in the metric system equal to 1000 kilograms, or 2205 pounds. (*See* **Ton**.)

Tor

A unit of pressure equal to one Nm^{-2}. It was suggested in 1913 but was seldom used[23]. It was named after E. Torricelli (1608–47), the Florentine scientist who invented the mercury thermometer.

Torr

A standard atmosphere (101 325 Nm^{-2}) has, by definition, a pressure of 760 torr. Thus the torr is equivalent to a pressure of one millimetre of mercury within one part in 7×10^6; it is also equal to 1·33322 millibars. The torr was adopted by the British Standards Institution[24] in 1958 but was used in Germany for several decades previous to this. The unit is named after E. Torricelli (1608–47), the Florentine scientist who in 1643 discovered the properties of the barometer.

Townsend

A CGS unit of electrical breakdown in a gas which is defined as E/N, where E is the field strength (V cm^{-1}) and N is the gas number (cm^{-3}); thus 1 townsend is equal to 10^{-17} V cm^2. The unit is named after Sir John Townsend[25] (1868–1957) who initiated the kinetic theory of ions and electrons in gases.

Transmission unit (T.U.)

The decibel was sometimes called a transmission unit[26] in the years 1923–28. Two powers P_1 and P_2 differ by N transmission units where N is defined by $N = 10 \log_{10} P_1/P_2$.

Trichromatic unit (T unit)

It is possible for a mixture of three primary colours to match white or any other colour. The colours are red, green and blue. If k lumens of a colour C be matched by l lumens of red, m of green and n of blue, then the colour or trichromatic equation is

$$k(C) \equiv l(R) + m(G) + n(B)$$

For a white match (W)

$$w(W) \equiv o(R) + p(G) + q(B)$$

where w, o, p and q have the same significance as k, l, m and n. If it is assumed that equal numbers of units of each of the primaries are required to match white, then the units in which the primaries are measured are no longer lumens but represent definite quantities of stimuli. In this case

$$k(C) \equiv \frac{l}{o}(R) + \frac{m}{p}(G) + \frac{n}{q}(B)$$

or

$$k(C) \equiv R(R) + G(G) + B(B)$$

The amount of colour represented by a trichromatic equation in which the sum of the coefficients of (R), (G) and (B) is unity is a trichromatic or T unit. Thus

$$\frac{k}{S}(C) \equiv \frac{R}{S}(R) + \frac{G}{S}(G) + \frac{B}{S}(B)$$

in which $S = R + G + B$

or

$$1\cdot0(C) \equiv r(R) + g(G) + b(B)$$

This equation is the unit trichromatic equation, r, g and b being the trichromatic coefficients. One T unit of (C) is equal to k/S lumens of (C) and if CT units of (C) are matched

$$C(C) \equiv R(R) + G(G) + B(B)$$

where $C = R + B$ and R, G and B are called the tristimulus values of the colour.

In 1931 the Committee Internationale d'Eclairage suggested[27] that instead of colours of actual physical stimuli three reference stimuli known as (X), (Y) and (Z) be used and defined them as

$$(X) = \quad 2\cdot3646(R) - 0\cdot5151(G) + 0\cdot0052(B)$$
$$(Y) = -0\cdot8965(R) + 1\cdot4264(G) - 0\cdot0144(B)$$
$$(Z) = -0\cdot4618(R) + 0\cdot0887(G) + \cdot10092(B)$$

where (R), (G) and (B) are monochromatic radiations of wavelengths 7000, 5461 and 4358 Å respectively (700·0; 546·1; 435·8 nm).

In 1945[28] the names Munsell (A. H. Munsell, 1858–1918), Young (Thomas Young, 1773–1829) and Ostwald (W. F. Ostwald, 1853–1932) were suggested for X, Y and Z but a year later Konig[29] (A. Konig, 1856–1901) was substituted for Munsell and Priest (I. G. Priest, 1886–1932) for Ostwald.

Troland (Trol)
The troland is the retinal illumination produced by a surface having a luminance of one candela per square metre when the area of the aperture of the eye is one square millimetre. It is named after L. T. Troland (1889–1932) who proposed[30] the unit under the name of photon in 1916. It is also called a luxon (q.v.).

Troy weights
See Appendix 3.

Turn over number
This is a measure of the catalytic power of an enzyme. It is the molecular activity defined as the number of molecules of substrate, or equivalents of the group concerned, transformed per minute by one molecule of enzyme at optimal substrate concentration. If the enzyme has a distinguishing prosthetic group or catalytic centre whose concentration can be measured, the turn over number is expressed as catalytic centre activity and is then defined as the number of molecules of substrate transformed per minute per

catalytic centre. Molecular activity equals the catalytic centre activity if the enzyme molecule contains one active centre: if there are n centres per enzyme molecule, the molecular activity will be n times the catalytic centre activity. It has been recommended by the International Biochemistry Committee that this unit be discontinued[31].

Twaddle
A specific gravity unit. (*See* **Degree (hydrometry)**.)

Typp
A unit used in the textile industry for describing the length per unit mass of a yarn. (*See* **Yarn counts**.)

U

U factor

The U factor describes the thermal performance of buildings[1]. It can be expressed in Btu ft^{-2} h^{-1} deg F^{-1} and should have a value of less than one. An 11-inch cavity brick wall has a U value of about 0·30. The SI unit is W m^{-2} °C^{-1}. Current building regulations in the UK specify a U factor of 0·6 W m^{-2} °C^{-1} as being desirable for roofs.

Units of activity

Units of activity are used to express the therapeutic properties of certain medical preparations (*see* **International biological standards**).

Universal time (UTC)

Time may be used either to specify the moment (or epoch) at which an event takes place, e.g. the battle of Hastings in 1066, or to describe a time interval, e.g. Brighton is an hour's journey by train from London. Today, the former can be derived from coordinated universal time (UTC) and the latter measured in SI seconds which are also used to denote time interval in UTC.

On 1 January 1972 UTC replaced Greenwich mean time (GMT) in scientific work and has subsequently been adopted as the legal standard of time in many countries[2–5]. The change was made because the atomic clock had become a better time keeper than the motion of the earth, the irregularities of which could alter the length of a year by nearly a second. The time indicated by an atomic clock is known as International Atomic Time (TAI) and is now used as a standard against which UTC is continuously compared. In 1972 UTC was deliberately set to be ten SI seconds behind TAI, i.e. TAI − UTS = 10 second. Then, as UTC varies, the gap between the two times can be adjusted by the addition or subtraction of whole seconds (leap seconds) so that all concerned can know by how much UTC is leading or lagging on TAI. Among the advantages of this system is that navigators, and others who used GMT, can continue to work from the same nautical almanacs as they had done previously for UTC is based on the movement of celestial bodies, and those interested in knowing the exact time at which an event has taken place can do so rapidly from TAI, whereas previously this could not be known with precision until after the completion of time consuming astronomical observations and their associated calculations.

V

Vac
A unit of pressure equal to 10^3 dyne cm^{-2}. It was suggested by Florescu in 1960 but there seems to be no reason for using the vac instead of the millibar[1] as both units are equal to 10^3 dyne cm^{-2}.

Var
A unit suggested[2] in 1954 for the magnitude of the reactive power in the equation.

$$\text{Power} = \text{Active power} + \text{Reactive power.}$$

Its name is derived from the initial letters of *v*olts, *a*mperes and *r*eactive power.

Verber
A name used in the early 1860s for the unit of charge. It was of the same order of magnitude as the coulomb.

Vickers pyramid number (VPN)
A unit of hardness. (*See* **Hardness numbers (solid)**.)

Vieth scale
A scale used in lactometry. (*See* **Degree (hydrometry)**.)

Violle
An alternative name for the candela (q.v.). It is named after J. L. G. Violle (1841–1923) who, in 1884, proposed that luminous intensity should be defined in terms of black body radiation.

Viscosity grades (ISO VG)
In 1975 the International Standards Organisation promulgated a viscosity grading for industrial lubricants[3]. The grades are chosen so that the kinematic viscosity of the mid-point of each grade is 50% higher than the preceding one, the viscosities covered by the scheme ranging from 2·2 cSt to 1500 cSt. A total of 18 grades are provided, all of which are described by a number which is approximately the same as the viscosity (expressed in centistokes) of the mid-point of grade concerned, e.g. ISO VG 2 represents a

151

mid-point viscosity of 2·2 cSt; ISO VG 3 represents 3·3 cSt; ISO VG 5 represents 4·6 cSt; . . . ISO VG 1500 represents 1500 cSt.

Viscosity index

The viscosity index is an empirical number indicating the effect of change of temperature on the viscosity of oil. A low viscosity index indicates a relatively large change of viscosity with temperature. The index, which can be negative or positive, is generally within the range ± 100. It was adopted in 1941 in the United States[4] and is also used in the United Kingdom.

Voegtlin

The unit of activity for pituitary extract (*see* **International biological standards**).

Volt (V)

The volt is the SI unit of electrical potential. It is equal to the difference in electrical potential between two points on a conductor carrying a constant currentof 1 ampere when the power dissipated between these points if 1 watt. The unit was first proposed in 1861[5] by Sir Charles Bright and Latimer Clark, who suggested the practical unit of potential should be based on the 'theoretical metrical unit', which should be multiplied by the necessary power of 10 to make the practical unit to be of the same order as the e.m.f. of a Daniell cell. The name suggested for the unit was the ohma. It was renamed the volt two years later after A. Volta (1745–1827), the Italian scientist who did pioneer work on electrical cells. The unit was adopted very slowly until 1873[6]; before this date most writers preferred to express potentials either in fundamental units or as some ratio of the e.m.f. developed by a Daniell or a Grove cell. The volt received international recognition in 1881[7], when it was one of the first practical units to be approved by the IEC. In 1948[8] the absolute volt replaced the international volt as the unit of potential, the latter unit having been introduced in 1908[9]. In practice the change is slight: whereas the e.m.f. of Weston cell at 20°C is 1·0183 international volts, its value in absolute units is 1·01859 volts. (1 international volt = 1·00034 absolute volts.)

Volume unit (vu)

A unit used in telecommunication engineering to describe the magnitude of complicated electric waves such as those associated with speech or music. The magnitude, or power, expressed in volume units, is equal numerically to the number of decibels by which it is greater or less than an arbitrary reference level. This level is the power produced by a sinusoidal wave developing one milliwatt in an impedance of 600 ohms. The unit was formally recognized by the Institute of Radio Engineers in 1938 in an attempt to have a standard system for measuring volume in telephony and allied sciences[10]. It was recognized by the American Standards Association in 1951[11].

W

Watt (W)
The watt is the SI and the practical unit of power. It is the power dissipated when 1 joule is expended in 1 second. The unit was proposed by C. W. Siemens in his presidential address to the British Association[1] in 1882. It is named after the Scottish engineer James Watt (1736–1819) who pioneered the development of the steam engine. The difficulty of using the letter W in the French language held up its international recognition until 1889. In 1908 the watt was defined in terms of the international ohm and the international ampere[2] but this international watt was replaced when the international units were superseded by the absolute units in 1948[3] (1 international watt = 1·00019 absolute watt).

Wave number
The name adopted by the British Association in 1872[4] to replace reciprocal wavelength in spectroscopy. In its original form the wave number represented the number of wavelengths per millimetre but today it is the number of wavelengths per centimetre. (*See* **Kayser**.) The name was proposed by G. J. Stoney and J. E. Reynolds in 1871[5].

Weber (electric current)
In 1881[6] the British Association proposed the practical unit of current should be the weber, but at the first meeting of the IEC, which was held shortly afterwards in Paris, it was decided to name the unit of current the ampere. This suggestion was readily followed by the British Association.

Weber (Wb) (magnetic flux)
The practical, as well as the SI unit of magnetic flux. It is equal to 10^8 maxwells, hence a change of flux at a uniform rate of 1 weber per second in a coil of N turns induces an e.m.f. of N practical volts in the coil. The unit was first mentioned by name by C. W. Siemens in his presidential address to the British Association in 1882[1]; it is named after W. E. Weber (1804–91),

one of the promoters of the absolute system of units. The unit was approved by the British Association in 1895[7], by the IEC[8] in 1933 and the 8th CGPM (1948) ruled that it would be the SI unit of magnetic flux density.

Weber (pole strength)

In the nineties the weber was given as the unit of pole strength in both Barker's *Advanced Physics* and in Latimer Clark's *Dictionary*[9]. A pole had a strength of 1 weber when it produced a field of 1 gauss (now called oersted) at a distance of 1 centimetre in air.

Weber number (We)

A dimensionless number[10] associated with surface tension waves in a manner analogous to the connection between the Froude number and gravity waves. The weber number is given by the expression $\rho V^2 l / \sigma$, where ρ is the density of the liquid, V the velocity of the wave, σ the surface tension and l has the dimensions of length.

Weisskopf unit

A unit expressing the transition probability of nuclei from one state to another. It was first used by V. F. Weisskopf in 1951[11] and was given a name in 1958[12]. It is nearly identical to the Moszkowski unit (q.v.).

Wet bulb globe temperature (WBGT) Heat stress index

A number used to express the heat stress imposed on the human body by a certain environment. It is given by the equation

$$\text{WBGT} = 0.7a + 0.2b + 0.1c$$

where a, b and c are the temperatures (in °C or °F) of the natural wet bulb, the standard globe and the surrounding air respectively. The wet bulb temperature is that shown by a normal mercury thermometer covered with a wet cotton sock. The standard globe temperature is that at the centre of a black copper sphere of 0.15 m diameter. For temperatures in °C, values of the index range from 18 to 33. About 40 suggestions for heat stress indices exist in the literature but the WBGT index[13] is the only one internationally standardized[14] (see ISO 7243).

Other stress indices which are in use are the effective temperature index[15] and the corrected effective temperature index, the predicted 4 hours sweat rate index[16], the heat stress index[17], the index of thermal stress[18], the new effective temperature index[19] and the required sweat rate index[20]. The latter is now being proposed by the ISO as a draft standard. A review of heat stress indices is given by Olesen[21].

Whizz

A unit proposed in 1981 by B. P. Kibble[22] of the National Physical Laboratory (UK) which, with the second and a redefined volt and ampere would give a system of units whose basis is completely free of material standards. Kibble proposes the whizz as 1/299 792 458 of the velocity of plane electro-magnetic radiation in free space. It is suggested that the volt could be that potential step which would correspond to radiation of $4.835\,94 \times 10^{14}$ Hz in a Josephson junction while the ampere could be the current flowing through a quantized Hall effect device which would give rise to a Hall potential of 6453·2 V. The kilogram, then a derived unit, could be that mass which, when moving with a velocity of one whizz, possesses mechanical energy equal to an electrical energy of 0·5 V-ampere-second.

Wink

The unit of time in a system of units based on the metre, gram and wink (1/3000 microsecond). In this system the unit of force is equal to 9×10^{13} N and is called a samson and that of work, an einstein (9×10^{13} J). Electrical units are also included with such names as the simon for the unit of resistance (30 ohms). The system was proposed apparently semi-facetiously in 1957[23].

Wire and plate gauges

The Standard Wire Gauge (SWG) is the legal wire gauge in the United Kingdom[24]. This gauge was sanctioned by the Board of Trade in 1883[25] at the request of engineers and manufacturers because at that time nearly every wire maker had his own gauge. These included an assortment of almost identical gauges known as the Birmingham wire gauges, namely the Wynn (1810), Cocker (1858), Mallock and Preece (1872), Walker (1878), Watkins (1878), the needle wire, the brass wire and the piano wire. The Standard Wire Gauge was legalized by Order in Council of 1893. Other gauges[26] in use were the South Staffordshire or Robinson (*c.* 1830), the Lancashire or Warrington devised by Peter Stubbs in 1843, the Rylands (1862), the Brown and Sharpe (1855), the Whitworth (1857), the Clark (1867), the Briggs (1877) and the Hughes (1878). The Birmingham Wire Gauge in use today, which is sometimes called the Stubbs iron wire gauge, was decided upon in 1884. In all gauges, except the Whitworth, the diameters get less as the gauge numbers increase. This follows a precedent established in the early eighteenth century, when the gauge number was reputed to indicate the number of times the wire had been drawn through the draw plate. Thus the wire would be thinner after the tenth draw (No. 10) than it would be after the ninth (No. 9). In the Whitworth gauge the number represented the diameter of the wire in mils, so that the numbers in this gauge increased as the diameters of the wire got greater.

In the USA the American Wire Gauge (AWG) is used. This originated in the gauge devised by Brown and Sharpe of Rhode Island in 1855. Like the Standard Wire Gauge it classifies wire diameters in geometrical progression. Each successive gauge number represents a wire with an area of cross-section 20 per cent less than its immediate predecessor. Other gauges used in the USA include the Steel wire gauge, the Washburn and the Moen, the Roebling, and the National wire gauges. The British Standard Wire Gauge is also employed but its use is mainly confined to telecommunication wires.

In France wire diameters are frequently described by the Paris gauge. This is a metric gauge with numbers from 1 to 30. No. 1 implies a thickness of 0·6 mm, No. 2 represents 0·7 mm, No. 10 is 1·5 mm and above this the scale opens out until No. 30 represents a diameter of 10 mm.

In the United Kingdom the thickness of plates may be indicated either by the Standard Wire Gauge or by the Birmingham Gauge (BG). The latter was adopted by the South Staffordshire Iron Masters in 1884 and was legalized by an Order in Council dated 16 July 1914[27]. The Steel Plate gauge, authorized in 1893[28], is used in the USA.

The six most common gauges in the English-speaking world are compared in Table 12. More detailed information can be obtained in the Machinery's *Handbook* or in Newnes' *Engineers' Handbook*. A history of wire gauges is given by Dickinson and Rogers[29].

TABLE 12. *Comparison of commonly used gauges. Thicknesses are given in inches and millimetres*

	Gauge number							
	0		1		20		36	
Name	in	mm	in	mm	in	mm	in	mm
SWG	0·324	8·23	0·300	7·62	0·0360	0·914	0·0076	0·193
BWG	0·340	8·64	0·300	7·62	0·0350	0·889	0·0040	0·102
AWG	0·325	8·26	0·289	7·34	0·0320	0·813	0·0050	0·127
Steel wire	0·307	7·80	0·283	7·19	0·0350	0·889	0·0090	0·229
BG (plate)	0·396	10·06	0·353	8·97	0·0390	0·991	0·0061	0·155
US steel plate	—	—	—	—	0·0359	0·912	0·0067	0·170

Wobbe index

The Wobbe index or number is a measure of the amount of heat released by a gas burner of constant orifice. It indicates the quality of the gas and is given by the expression $V\rho^{-1/2}$, where V is the gross calorific value in Btu per cubic foot at s.t.p. and ρ is the specific gravity. Methane, natural gas and town gas have values of about 1340, 1300 and 730 respectively. For convenience town gases are grouped according to their Wobbe indices each group being designated by the letter G followed by a number, thus:

G number	G4	G5	G6	G7	G8
Wobbe index	701–760	641–700	591–640	531–590	below 531

Wood screws

The standard description of a wood screw gives its material, the style of its head, its length and its diameter. The length may be given in inches or metric units, but the diameter is expressed as a number. Most countries have their own systems of numbers which run in the reverse direction to those used for wire gauges; i.e. in wood screws low numbers represent small diameters. Wood screw diameters have been standardized in Great Britain since about 1880[30]. Some wood screw standard diameters are given in Table[13].

TABLE 13. *Comparison of commonly used wood screw sizes. Diameters are given in inches and millimetres*

Size	British		American		French		German	
	in	*mm*	*in*	*mm*	*in*	*mm*	*in*	*mm*
0	0·060	1·524	0·058	1·473	—	—	0·065	1·651
10	0·192	4·880	0·189	4·800	0·057	1·448	0·181	4·600
20	0·332	8·430	0·321	8·150	0·157	3·990	0·339	8·610
30	0·472	11·990	0·482	12·240	0·413	10·490	—	—

X

X unit (XU)
A unit equal to approximately 10^{-13} m used for X-ray wave lengths. (*See*
Siegbahn unit.)

Y

Yard (yd)

The yard is the fundamental unit of length in the foot pound second system of units. It is a very old measure which has maintained approximately the same value for over 800 years, the yard of Henry I (1100–37) being within a tenth of an inch of the yard in use today. In Britain until 1963 the imperial standard yard was the legal unit of length. It was defined in the Weights and Measures Act of 1878 as the distance between two points on a specified metal bar[1]. In the 1963 Weights and Measures Act[2] it was redefined as being equal to 0·9144 metres exactly, a definition which made the inch equal to exactly 2·54 cm. Four years before this, however, the 0·9144 metre yard had been approved for scientific purposes in the British Commonwealth and the United States of America, being referred to as the international yard[3].

In the United States the yard was specified by the Mendenhall Order of 1893 as being equal to 3600/3937 meter, this made the US yard 0·91440183 metre, i.e. about $0·7 \times 10^{-4}$ inch longer than the British 1963 yard.

Some archaeologists, from a statistical study of the dimensions of prehistoric stone circles, claim to have found evidence of the use of a unit of length of 2·72 ft (0·839 m) which they call the megalithic yard[4].

Yarn counts

In the textile industry the yarn count or yarn number gives either the mass per unit length or the length per unit mass of a yarn fibre[5]. Since 1950 metric units of mass and length have been used. In the western hemisphere the unit is the drex, which gives the mass in grams per 10 000 metre length of yarn. The drex was proposed in 1945 and was recognized in 1949. In the Eastern hemisphere, the tex, which is the mass in grams per 1000 metres, is used. This unit was adopted by most European countries over the years 1947 to 1951, Spain being the first country to recognize it. All other system of yarn counts are now technically obsolete, but it will be a long time before they go completely out of use. Thus at present denier is still widely used. A list of the more important yarn count units, together with the value equivalent to 20 tex, are given in Table 14.

TABLE 14. *Yarn count units*[6]

Name	Unit of mass	Unit of length	Value equal to 20 tex
Direct (mass per unit length)			
denier[a]	gram	9000 m	180
drex[b]	gram	10 000 m	200
—	grain	100 yd	28·2
—	grain	120 yd	33·9
poumar[c]	pound	10^6 yd	40·3
tex	gram	1000 m	20
Indirect (length per unit mass)			
American run	ounce	100 yd	15·5
cotton	pound	840 yd (hand)	29·5
linen	pound	300 yd (lea)	82·7
metric	kg	1000 m	50
typp[d]	pound	1000 yd	24·8
worsted	pound	560 yd	44·3

[a] Introduced Paris 1900
[b] In use in Canada and USA
[c] *Pounds per million yards*
[d] *Thousand yards per pound*

Year (a)

A year is the time required for the earth to go round the sun. Its approximate value of $365\frac{1}{4}$ days was known to the Egyptians and Babylonians over four thousand years ago, and today it is often referred to as a Julian year.

There are a number of year units derived from the orbital movement of the earth. Since to an observer on the earth this movement gives rise to the impression that the sun is moving in a circle, called the ecliptic, on a sphere with the earth as centre, known as the celestial sphere, the units are commonly defined with respect to the sun's apparent motion. The ecliptic cuts the celestial equator at two points, one of which is the First Point of Aries on the vernal equinox, i.e. the noon position of the sun as it crosses the equator in March. A solar or tropical year (a_{trop}) is the average time for the sun to return to the First Point of Aries. The 315 569 259 747th part of the tropical year at 1900 January 0, 12 noon Ephemeris time (Dec. 31, 1899, 12 noon Universal time) is the Ephemeris second. A sidereal year (a_{sid}) is the average time for the earth to complete one revolution round the sun with reference to the fixed stars. An anomalistic year (a_{anom}) is the interval between successive passages of the sun through the perigee. An astronomical year (a_{astr}), also known as the Besselian year or annus fictus, is the time during which the right ascension of the mean sun, i.e. the angle between the meridians of the sun and the vernal equinox, increases by 360°. If the year is measured from the instant the sun's meridian is 280°, the year begins on January 1 and this is the civil year. A Gaussian year (a_{gauss}) is the theoretical time, calculated from Kepler's laws, for the earth to go round the sun.

The durations in days for the different years are:

a_{trop}	$365{\cdot}242\,199 - 0{\cdot}000\,006\ T$
a_{astr}	$365{\cdot}253\,189\,7$
a_{sid}	$365{\cdot}256\,360 + 0{\cdot}000\,000\,1\ T$
a_{anom}	$365{\cdot}256\,641 + 0{\cdot}000\,003\ T$
a_{gauss}	$365{\cdot}258\,898$

where T is the time in centuries from 1 January 1900.

$$a_{sid} - a_{trop} = 20{\cdot}4\ \text{min}; \qquad a_{anom} - a_{trop} = 20{\cdot}8\ \text{min}$$

Young

A name suggested in 1945 for the Y stimulus in the trichromatic colour system (*see* **Trichromatic unit**). It is named after T. Young (1773–1829) who established the wave theory of light.

Z

Zhubov scale (ball)

This scale is used in Russia to report ice conditions[1] in much the same way as wind velocities are described on the Beaufort scale. The scale goes from one ball to ten, thus one ball indicates that about 10% of the visible surface is covered with ice, 2 balls denotes 20% coverage, etc. In general:

1–3 (ball)	implies sparse ice cover offering little resistance to shipping,
4–6 (ball)	broken ice, speed should be reduced by about a quarter,
7–8 (ball)	concentrated ice, ice-breaker may be required,
9–10 (ball)	complete cover, navigation dependent on type of ice and effectiveness of ice-breakers.

Appendices

1. Fundamental physical constants

(Scale $C^{12} = 12.000$ amu)

Acceleration of gravity	g	$9.812\,60$ m s^{-2}
Atomic mass unit	u	$1.660\,565\,5(86) \times 10^{-27}$ kg
Avogadro constant	N_A	$602.204\,5(31) \times 10^{21}$ mol^{-1}
Boltzmann constant	k	$13.806\,62(44) \times 10^{-24}$ J K^{-1}
Compton wavelength of an electron	λ_c	$2.426\,308\,9(40) \times 10^{-12}$ m
Compton wavelength of a proton	λ_{cp}	$1.321\,409\,9(22) \times 10^{-15}$ m
Electron, charge	e	$160.218\,92(46) \times 10^{-21}$ C
Electron, rest mass	m	$910.953\,4(47) \times 10^{-33}$ kg
Electron, ratio charge/mass	e/m	$175.880\,47(49) \times 10^9$ C kg^{-1}
Electron, magnetic moment (Bohr magneton)	μ_B	$9.274\,078(36) \times 10^{-24}$ J T^{-1}
Faraday constant	F	$96.484\,56(27) \times 10^3$ C mol^{-1}
Fine structure constant	α	$0.007\,297\,350\,6(60)$
First Bohr radius	a_0	$52.917\,706(44) \times 10^{-12}$ m
Gravitational constant	G	$66.720(41) \times 10^{12}$ N m^2 kg^{-2}
Josephson frequency/voltage ratio	$2e/h$	$483.593\,9(13) \times 10^{12}$ Hz V^{-1}
Magneton, nuclear	μ_N	$5.050\,824(20) \times 10^{-27}$ J T^{-1}
Mass energy conversion factor	$-$	$931.501\,6(26)$ MeV
Mechanical equivalent (15° calorie)	J	$4.185\,5\,(55)$ J cal^{-1}
Muon, rest mass	m_μ	$188.356\,6(11) \times 10^{-30}$ kg
Neutron rest mass	m_n	$1.674\,954\,3(86) \times 10^{-27}$ kg
Permeability of free space	μ_0	$4\pi \times 10^{-7}$ H m^{-1}
		$1.256\,637\,061\,4 \times 10^{-6}$ H m^{-1}
Permittivity of free space	ε_0	$8.854\,187\,818 \times 10^{-12}$ F m^{-1}
Planck constant	h	$662.617\,6(36) \times 10^{-36}$ J s
Proton rest mass	m_p	$1.672\,648\,5(86) \times 10^{-27}$ kg
Rydberg constant	R	$10.973\,731\,77(83) \times 10^6$ m^{-1}
Speed of light	c	$299.792\,458 \times 10^6$ m s^{-1}
Standard volume	V_0	$22.413\,83(70) \times 10^{-3}$ m^3 mol^{-1}
Stefan-Boltzmann constant	σ	$56.703\,2(71) \times 10^{-9}$ W m^{-2} K^4
Triple point of water	$-$	273.16 K
Universal gas constant	R	$8.314\,41(26)$ J mol^{-1} K^{-1}

Voltage–wavelength conversion factor   – $1 \cdot 237\,282\,0(66) \times 10^{-6}$ m V^{-1}
kX unit/ångström conversion factor Λ $1 \cdot 002\,077\,2(54)$

The numbers in parentheses are the standard deviation uncertainty in the last digits of the quoted value.

Cohen, E. R., *Atomic Data and Nuclear Data Tables (USA)*, **18**, 587 (1976).
Cohen, E. R. and Taylor, N. B., *J.Phys. and Chem. Ref. Data (USA)*, **2**, 663 (1973).

More precise values are being considered by the Task Group on Fundamental Constants of the Committee on Data for Science and Technology (CODATA). See *Precision Measurements and Fundamental Constants II.* (Eds: N. B. Taylor and W. D. Phillips.) Natl. Buv. Stand. (US) Spec. Publ. **617** (1984).

2. Standardization committees and conferences

The definition and recommendation of the use of units has been carried out by a number of bodies, usually of international standing. The most prominent of these are the Conférence Internationale des Poids et Mesures (CIPM), the International Electrotechnical Commission (IEC), the Commission International de l'Eclairage (CIE), the International Union of Pure and Applied Physics (IUPAP) and the International Union of Pure and Applied Chemistry (IUPAC). Other standardization committees are the International Organization for Standardization, the British Standards Institution and the American Standards Association. In Great Britain during the nineteenth and early part of the twentieth century the British Association for the Advancement of Science played a prominent part in proposing and establishing units. Units used in commerce are generally defined either by Acts of Parliament, by Orders in Council or by Board of Trade Orders. In the United States commercial units are legalized by Acts of Congress.

CGPM (Conférence Générale des Poids et Mesures)
CIPM (Commission International des Poids et Mesures)
The Conférence Générale des Poids et Mesures (CGPM) was the direct outcome of an international meeting held in Paris in 1875. Its object was the promotion of international uniformity in standards of measurement and it resulted in the signing of a treaty, known as the Convention du Metre, by 21 nations. The terms of the treaty included the setting up of a permanent institution known as the Bureau International des Poids et Mesures at Sèvres, near Paris, which would be managed by the Commission International des Poids et Mesures (CIPM) on behalf of the CGPM. The Commission consists of 18 members, it meets annually and is responsible for preparing the agenda for the full conference which, in recent times, has met at four yearly intervals. The Commission has also set up several specialist advisory committees to assist it in the planning of its research and the

drawing up of recommendations for definitions of units of measurement. Initially the CGPM was concerned only with the measurement of mass and length but today it has the oversight of all scientific measurements ranging from photometry to time.

ICRU (International Commission on Radiological Units)
This organization was formed in 1925. It meets at regular intervals.

IEC (International Electrotechnical Commission)
The International Electrotechnical Commission first met in Paris in 1881. At this meeting the electromagnetic CGS units were adopted as the fundamental units of electricity and magnetism and the names ampere, coulomb, farad, ohm and volt were agreed for five practical units. The British delegation withdrew their proposal to call the unit of current the weber so that the names of two French scientists, A. M. Ampère (1775–1836) and C. A. Coulomb (1736–1806) could be associated with the first internationally accepted electrical units. The custom of naming any new units after eminent scientists of the country in which the IEC meets is still continued; thus at the Chicago meeting in 1893 the henry (Joseph Henry, 1797–1878) was adopted for the practical unit of induction and at Oslo in 1930 the oersted (H. C. Oersted, 1777–1851) was chosen as the name for the unit of magnetizing force.

CIE (Commission International de l'Eclairage)
This international body considers the standards and definitions used in photometry and calorimetry. It has met at intervals in different parts of the world since its inauguration in Paris in July 1921. It is frequently referred to as the CIE or the ICI, the latter being the initials of the English form of its name – International Commission for Illumination.

IUPAP (International Union of Pure and Applied Physics)
This organization set up a committee known as the SUN (Standardization, Units and Nomenclature) Committee in July 1931 to consider the definitions and names of physical units. The committee meets at suitable intervals. One of its first recommendations was that the unit of heat, when measured in terms of energy, should be the joule.

IUPAC (International Union of Pure and Applied Chemistry)
One of the objects of this body is the establishment of a standard nomenclature for chemistry. Standard nomenclature was first suggested by T. O. Bergmann (1735–84) in 1770 and re-introduced by J. J. Berzelius (1779–1848) in 1813. In 1892 an International Conference on Nomenclature was held in Geneva which produced a semblance of order in the names used

in chemistry. This conference continued to meet at irregular intervals and formed the nucleus of the present International Union of Pure and Applied Chemistry which held its first meeting in 1919. Two years later nomenclature committees were set up. Recent reports issued by these committees include the Commission on Nomenclature of Inorganic Chemistry (1957) and in the same year one on Organic Chemistry. It is interesting to note that after two hundred years of endeavour to have a standard nomenclature, element 41 is called columbium in the USA and niobium in Great Britain.

ISO (International Organization for Standardization) Geneva

This was set up in 1926; its first president was Sir Richard Glazebrook who was Director of the National Physical Laboratory at Teddington, Middlesex. The organization publishes recommendations which may be adopted as national standards by the individual member countries.

BIH (Bureau Internationale de l'Heure)

This is the international organization responsible for the coordination of time throughout the world. The Bureau is situated in Paris and was established by the International Time Conference which was held in October 1912.

BSI and ASA (British Standards Institution and American Standards Association)

Many countries have their own independent standards associations; thus in Great Britain the British Standards Institution, founded in 1901 as the British Engineering Standards Association, issues recommendations known as British Standards as to the sizes and the names of units. Today there are over three thousand standards in current use, all of which are numbered and carry the initial letters BS, followed by the number and date of the publication or standard.

In the USA the American Standards Association (ASA), founded in 1917, carries out a similar function.

3. Tables of weights and measures

(For comprehensive list see *Encyclopaedia Britannica*)

UNITS OF CAPACITY

In the UK the commercial units of capacity both for fluids and dry substances are the gallon and units derived from it. In the US the gallon and derived units are legal measures only for fluids; for dry substances, the units are the bushel and units derived from it.

UK (liquids and solids)

60 minims	= 1 fluid drachm
8 fluid drachms	= 1 fluid ounce
5 fluid ounces	= 1 gill
4 gills	= 1 pint
2 pints	= 1 quart
4 quarts	= 1 gallon
2 gallons	= 1 peck
4 pecks	= 1 bushel
3 bushels	= 1 sack
8 bushels	= 1 quarter
12 sacks	= 1 chaldron
1 UK gallon	= 1·20094 US gallon = 277·42 in^3
1 UK minim	= 0·960754 US minim

US (liquid)

60 minims	= 1 fluid dram
8 fluid drams	= 1 fluid ounce
4 fluid ounces	= 1 gill

174

4 gills	= 1 liquid pint
2 liquid pints	= 1 liquid quart
4 liquid quarts	= 1 gallon
42 gallons	= 1 barrel
1 US gallon	= 231 in^3

US (dry measure)

2 dry pints	= 1 dry quart
8 dry quarts	= 1 peck
4 pecks	= 1 bushel
1 bushel (US)	= 0.968 939 bushel (UK)
	= 2150·42 in^3 = 35·239 071 669 dm^3

The bushel is sometimes called the stricken or struck bushel.

UNITS OF AREA

144 square inches	= 1 square foot
9 square feet	= 1 square yard
$30\frac{1}{4}$ square yards	= 1 square rod, pole or perch
40 square rods	= 1 rood
4 roods	= 1 acre
640 acres	= 1 square mile
4840 square yards	= 1 acre
1 square yard	= 0·836127 square metre
1 acre	= 2·25 vergees (Jersey)
	= 2·625 vergees (Guernsey)

UNITS OF LENGTH

12 lines	= 1 inch (in)
12 inches	= 1 foot (ft)
3 feet	= 1 yard (yd)
$5\frac{1}{2}$ yards	= 1 rod, pole or perch
40 rods	= 1 furlong
8 furlongs	= 1 mile

The English mile is 1760 yards whereas the Irish mile (obsolete) is 2240 yards. A Gunter's chain consists of a hundred links and is 22 yards long. In the USA an Engineer's chain of 100 feet is also used.

1 yard	= 0·9144 metre
5 miles	≃ 8 kilometres

CLOTH MEASURE

$2\frac{1}{4}$ ins = 1 nail
4 nails = 1 quarter yard

UNITS OF MASS

Avoirdupois weights

The avoirdupois system of weights is the legal system for commerce in most English-speaking countries. The fundamental unit of the system is the pound, which is equal to 7000 grains. The mass of the pound has remained consistent to within 0·1 per cent since when it was first defined in 1340. The avoirdupois system, like the pound, has changed little since the fourteenth century. From time to time the legal definitions of some of the weights have been more precisely described by Act of Parliament, the last major revision being authorized in the Weights and Measures Act of 1963.

16 drams (dr) = 1 ounce (oz)
16 ounces (oz) = 1 pound (lb)
14 pounds (lb) = 1 stone
28 pounds (lb) = 1 quarter
112 pounds (lb) = 1 hundredweight (cwt)
2240 pounds (lb) = 1 ton

In the USA 100 lb is called a short hundredweight (short cwt) and the 112 lb hundredweight is the long hundredweight. The USA or short ton is equal to 2000 pounds. In both the USA and the UK the pound is equal to 7000 grains.

(1 pound≃0·453 592 kilogram)

Apothecaries weights

Apothecaries weights were at one time used in pharmacy. In this system the ounce had the same mass as the Troy ounce – 480 grains. They were legalized as a consequence of the Medical Education Acts of 1858 and 1862. The weights are now obsolete and have been removed from recent editions of the British Pharmacopoeia.

20 grains = 1 scruple
3 scruples = 1 drachm
24 scruples = 1 ounce (oz apoth.)

In the USA the drachm is called a dram and the abbreviation for the ounce is oz ap.

Troy weights

The Troy system of weights was at one time the legal system for the weighing of precious metals in the United Kingdom. The fundamental unit was the Troy pound, which was equal to 5760 grains. Troy weights were introduced in England in the early fifteenth century. Although used extensively for the weighing of precious metals the system was not officially legalized until 1824. The Troy pound was abolished in 1878 but the Troy ounce of 480 grains is still a legal unit of mass. Tradition maintains the name Troy is derived from Troyes, the French town famous for its great medieval fairs.

$$24 \text{ grains} = 1 \text{ pennyweight (dwt)}$$
$$20 \text{ pennyweight} = 1 \text{ ounce}$$
$$12 \text{ ounces} = 1 \text{ pound}$$

Troy units are used for weighing precious metals. Troy units are no longer legal in the UK but they are still legal in the USA.

(The troy pound weighs 5760 grains or 0·3732 kilogram)

Hay and straw weight

$$36 \text{ lbs straw} = 56 \text{ lbs Old Hay}$$
$$= 60 \text{ lbs New Hay}$$
$$= 1 \text{ truss}$$
$$36 \text{ trusses} = 1 \text{ load}$$

MIXED UNITS

Barrels

1 barrel of alcohol	= 50 gallons (US)	= 189 litres
1 barrel of petroleum	= 42 gallons (US)	= 159 litres
1 barrel of salt	= 280 lb	= 127 kg
1 barrel of cement	= 376 lb	= 171 kg

Beer, wine and spirit measures

The following measures are frequently used in the wine and spirit trade. The values vary with the wine and spirit concerned; thus a hogshead of brandy can lie between 56 and 61 gallons, and the same measure of Madeira from 45 to 48 gallons. All measurements given below are expressed in imperial gallons:

Aum (hock)	30–32
Butt	108–117
Hogshead	44–60
Octave (whisky)	16

Pipe	90–120
Puncheon	90–120
Quarter	17–30
Stuck (hock)	260–265
Tonneau or tum	190–200

There are six bottles of wine to an imperial gallon; each bottle therefore contains 1 reputed quart.

2 reputed quarts	= 1 magnum
4 ,, ,,	= 1 jeroboam
6 ,, ,,	= 1 rehoboam
9 ,, ,,	= 1 methuselah
12 ,, ,,	= 1 salmarazd
16 ,, ,,	= 1 belshazzar
20 ,, ,,	= 1 nebuchadnezzar

Beer

$4\frac{1}{2}$ gallons = 1 pin	3 puncheons = 1 tun
2 pins = 1 firkin	6 firkins = 1 hogshead
4 firkins = 1 barrel	2 hogsheads = 1 butt
2 barrels = 1 puncheon	2 butts = 1 ton

In Ireland a pin is equal to 4 gallons.

4 Conversion tables

1. Length

		metre	inch	foot	yard	rod	mile
1 metre	=	1	39·37	3·281	1·093	0·1988	$6·214 \times 10^{-4}$
1 inch	=	$2·54 \times 10^{-2}$	1	0·083	0·02778	$5·050 \times 10^{-3}$	$1·578 \times 10^{-5}$
1 foot	=	0·3048	12	1	0·3333	0·0606	$1·894 \times 10^{-4}$
1 yard	=	0·9144	36	3	1	0·1818	$5·682 \times 10^{-4}$
1 rod	=	5·029	198	16·5	5·5	1	$3·125 \times 10^{-3}$
1 mile	=	1609	63360	5280	1760	320	1

1 Imperial standard yard = 0·914 398 41 metre
1 yard (Scientific) = 0·9144 metre
1 US yard = 0·914 401 83 metre
1 English nautical mile = 6080 ft = 1853·18 metres
1 International nautical mile = 1852 metres = 6076·12 ft

2. Area

	sq. metre	sq. inch	sq. foot	sq. yard	acre	sq. mile
1 sq. metre =	1	1550	10·76	1·196	$2·471 \times 10^{-4}$	$3·861 \times 10^{-7}$
1 sq. inch =	$6·452 \times 10^{-4}$	1	$6·944 \times 10^{-3}$	$7·716 \times 10^{-4}$	$1·594 \times 10^{-7}$	$2·491 \times 10^{-10}$
1 sq. foot =	0·0929	144	1	0·1111	$2·296 \times 10^{-5}$	$3·587 \times 10^{-8}$
1 sq. yard =	0·8361	1296	9	1	$2·066 \times 10^{-4}$	$3·228 \times 10^{-7}$
1 acre =	$4·047 \times 10^{3}$	$6·273 \times 10^{6}$	$4·355 \times 10^{4}$	4840	1	$1·563 \times 10^{-3}$
1 sq. mile =	$259·0 \times 10^{4}$	$4·015 \times 10^{9}$	$2·788 \times 10^{7}$	$3·098 \times 10^{6}$	640	1

1 arc = 100 sq. metres = 0·01 hectare
1 circular mil = $5·067 \times 10^{-10}$ sq. metre
$= 7·854 \times 10^{-7}$ sq. in
1 acre $= 0·4047$ hectare

3. Volume

		cubic metre	cubic inch	cubic foot	UK gallon	US gallon
1 cubic metre	=	1	6.102×10^4	35.31	220.0	264.2
1 cubic in	=	1.639×10^{-5}	1	5.787×10^{-4}	3.605×10^{-3}	4.329×10^{-3}
1 cubic ft	=	2.832×10^{-2}	1728	1	6.229	7.480
1 UK gallon*	=	4.546×10^{-3}	277.4	0.1605	1	1.201
1 US gallon†	=	3.785×10^{-3}	231.0	0.1337	0.8327	1

*Volume of 10 lb of water at 62°F.
†Volume of 8.328 28 lb of water at 60°F.
1 cubic metre = 999.972 litres
1 litre = 1000.028 cm^3 (volume of 1 kg of water at maximum density)

4. Angle

		degree	minute	second	radian	revolution
1 degree	=	1	60	3600	$1 \cdot 745 \times 10^{-2}$	$2 \cdot 778 \times 10^{-3}$
1 minute	=	$1 \cdot 667 \times 10^{-2}$	1	60	$2 \cdot 909 \times 10^{-4}$	$4 \cdot 630 \times 10^{-5}$
1 second	=	$2 \cdot 778 \times 10^{-4}$	$1 \cdot 667 \times 10^{-2}$	1	$4 \cdot 848 \times 10^{-6}$	$7 \cdot 716 \times 10^{-7}$
1 radian	=	$57 \cdot 30$	3438	$2 \cdot 063 \times 10^{5}$	1	$0 \cdot 1592$
1 revolution	=	360	$2 \cdot 16 \times 10^{4}$	$1 \cdot 296 \times 10^{6}$	$6 \cdot 283$	1

$1 \text{ mil} = 10^{-3} \text{ radian}$

5. Time

		year	solar day	hour	minute	second
1 year	=	1	365·24*	$8·766 \times 10^{3}$	$5·259 \times 10^{5}$	$3·156 \times 10^{7}$
1 solar day	=	$2·738 \times 10^{-3}$	1	24	1440	$8·640 \times 10^{4}$
1 hour	=	$1·141 \times 10^{-4}$	$4·167 \times 10^{-2}$	1	60	3600
1 minute	=	$1·901 \times 10^{-6}$	$6·944 \times 10^{-4}$	$1·667 \times 10^{-2}$	1	60
1 second	=	$3·169 \times 10^{-8}$	$1·157 \times 10^{-5}$	$2·778 \times 10^{-4}$	$1·667 \times 10^{-2}$	1

1 year = 366·24 sidereal days
1 sidereal day = 86 164·090 6 seconds
*Exact figure = 365·242 192 64 in 2000 A.D.

6. Mass

		kg	lb	$slug$	$metric\ slug$	$UK\ ton$	$US\ ton$	amu
1 kg	=	1	2·205	$6·852 \times 10^{-2}$	0·1020	$9·842 \times 10^{-4}$	$11·02 \times 10^{-4}$	$6·024 \times 10^{26}$
1 lb	=	0·4536	1	$3·108 \times 10^{-2}$	$4·625 \times 10^{-2}$	$4·464 \times 10^{-4}$	$5·000 \times 10^{-4}$	$2·732 \times 10^{26}$
1 slug	=	14·59	32·17	1	1·488	$1·436 \times 10^{-2}$	$1·609 \times 10^{-2}$	$8·789 \times 10^{27}$
1 metric slug	=	9·806	21·62	0·6720	1	$9·652 \times 10^{-3}$	$1·081 \times 10^{-2}$	$5·907 \times 10^{27}$
1 UK ton	=	1016	2240	69·62	103·6	1	1·12	$6·121 \times 10^{29}$
1 US ton	=	907·2	2000	62·16	92·51	0·8929	1	$5·465 \times 10^{29}$
1 amu	=	$1·660 \times 10^{-27}$	$3·660 \times 10^{-27}$	$1·137 \times 10^{-28}$	$1·693 \times 10^{-28}$	$1·634 \times 10^{-30}$	$1·829 \times 10^{-30}$	1

Imperial standard pound = 0·453 592 338 kilogram
US pound = 0·453 592 427 7 kilogram
International pound = 0·453 592 37 kilogram
1 tonne = 10^3 kilogram

7. Force

		dyne	newton	pound force	poundal	gram force
1 dyne	=	1	10^{-5}	$2 \cdot 248 \times 10^{-6}$	$7 \cdot 233 \times 10^{-5}$	$1 \cdot 020 \times 10^{-3}$
1 newton	=	10^{5}	1	$0 \cdot 2248$	$7 \cdot 233$	$102 \cdot 0$
1 pound force	=	$4 \cdot 448 \times 10^{5}$	$4 \cdot 448$	1	$32 \cdot 17$	$453 \cdot 6$
1 poundal	=	$1 \cdot 383 \times 10^{4}$	$0 \cdot 1383$	$3 \cdot 108 \times 10^{-2}$	1	$14 \cdot 10$
1 gram force	=	$980 \cdot 7$	$980 \cdot 7 \times 10^{-5}$	$2 \cdot 205 \times 10^{-3}$	$7 \cdot 093 \times 10^{-2}$	1

8. Power

		Btu per hr	ft lb s^{-1}	kg metre s^{-1}	cal s^{-1}	HP	watt
1 Btu per hour	=	1	$0 \cdot 2161$	$2 \cdot 987 \times 10^{-2}$	$6 \cdot 999 \times 10^{-2}$	$3 \cdot 929 \times 10^{-4}$	$0 \cdot 2931$
1 ft lb per second	=	$4 \cdot 628$	1	$0 \cdot 1383$	$0 \cdot 3239$	$1 \cdot 818 \times 10^{-3}$	$1 \cdot 356$
1 kg metre per second	=	$33 \cdot 47$	$7 \cdot 233$	1	$2 \cdot 343$	$1 \cdot 315 \times 10^{-2}$	$9 \cdot 807$
1 cal per second	=	$14 \cdot 29$	$3 \cdot 087$	$4 \cdot 268 \times 10^{-1}$	1	$5 \cdot 613 \times 10^{-3}$	$4 \cdot 187$
1 HP	=	2545	550	$76 \cdot 04$	$178 \cdot 2$	1	$745 \cdot 7$
1 watt	=	$3 \cdot 413$	$0 \cdot 7376$	$0 \cdot 1020$	$0 \cdot 2388$	$1 \cdot 341 \times 10^{-3}$	1

1 International watt = 1·000 19 absolute watt

9. Energy, work, heat

	Btu	joule	ft lb	cm^{-1}	cal	kW h	electron volt	kg^*	amu^*
1 Btu =	1	$1 \cdot 055 \times 10^{3}$	$778 \cdot 2$	$5 \cdot 312 \times 10^{25}$	252	$2 \cdot 930 \times 10^{-4}$	$6 \cdot 585 \times 10^{21}$	$1 \cdot 174 \times 10^{-14}$	$7 \cdot 074 \times 10^{12}$
1 joule =	$9 \cdot 481 \times 10^{-4}$	1	$7 \cdot 376 \times 10^{-1}$	$5 \cdot 035 \times 10^{22}$	$2 \cdot 389 \times 10^{-1}$	$2 \cdot 778 \times 10^{-7}$	$6 \cdot 242 \times 10^{18}$	$1 \cdot 113 \times 10^{-17}$	$6 \cdot 705 \times 10^{9}$
1 ft lb =	$1 \cdot 285 \times 10^{-3}$	$1 \cdot 356$	1	$6 \cdot 828 \times 10^{22}$	$3 \cdot 239 \times 10^{-1}$	$3 \cdot 766 \times 10^{-7}$	$8 \cdot 464 \times 10^{18}$	$1 \cdot 507 \times 10^{-17}$	$9 \cdot 092 \times 10^{9}$
1 cm^{-1} =	$1 \cdot 883 \times 10^{-26}$	$1 \cdot 986 \times 10^{-23}$	$1 \cdot 465 \times 10^{-23}$	1	$4 \cdot 745 \times 10^{-24}$	$5 \cdot 517 \times 10^{-30}$	$1 \cdot 240 \times 10^{-4}$	$2 \cdot 210 \times 10^{-40}$	$1 \cdot 332 \times 10^{-13}$
1 cal 15°C =	$3 \cdot 968 \times 10^{-3}$	$4 \cdot 187$	$3 \cdot 088$	$2 \cdot 108 \times 10^{23}$	1	$1 \cdot 163 \times 10^{-6}$	$2 \cdot 613 \times 10^{19}$	$4 \cdot 659 \times 10^{-17}$	$2 \cdot 807 \times 10^{10}$
1 kW h =	3412	$3 \cdot 600 \times 10^{6}$	$2 \cdot 655 \times 10^{6}$	$1 \cdot 813 \times 10^{29}$	$8 \cdot 598 \times 10^{5}$	1	$2 \cdot 247 \times 10^{25}$	$4 \cdot 007 \times 10^{-11}$	$2 \cdot 414 \times 10^{16}$
1 electron volt =	$1 \cdot 519 \times 10^{-22}$	$1 \cdot 602 \times 10^{-19}$	$1 \cdot 182 \times 10^{-19}$	$8 \cdot 066 \times 10^{3}$	$3 \cdot 827 \times 10^{-20}$	$4 \cdot 450 \times 10^{-26}$	1	$1 \cdot 783 \times 10^{-36}$	$1 \cdot 074 \times 10^{-9}$
1 kg^* =	$8 \cdot 521 \times 10^{13}$	$8 \cdot 987 \times 10^{16}$	$6 \cdot 629 \times 10^{16}$	$4 \cdot 525 \times 10^{39}$	$2 \cdot 147 \times 10^{16}$	$2 \cdot 497 \times 10^{10}$	$5 \cdot 610 \times 10^{35}$	1	$6 \cdot 025 \times 10^{2}$
1 amu^* =	$1 \cdot 415 \times 10^{-13}$	$1 \cdot 492 \times 10^{-10}$	$1 \cdot 100 \times 10^{-10}$	$7 \cdot 513 \times 10^{12}$	$3 \cdot 564 \times 10^{-11}$	$4 \cdot 145 \times 10^{-17}$	$9 \cdot 31 \times 10^{8}$	$1 \cdot 660 \times 10^{-27}$	1

*From the mass energy relationship $E = mc^2$

10. Pressure

	standard atmosphere	kg force cm^{-2}	dyne cm^{-2}	pascal	pound force in^{-2}	pound force ft^{-2}	millibar	torr	barometric in. Hg
1 standard atmosphere =	1	1·033	$1·013 \times 10^6$	$1·013 \times 10^5$	14·70	2116	1013	760	29·92
1 kg force cm^{-2} =	0·9678	1	$9·804 \times 10^5$	$9·804 \times 10^4$	14·22	2048	980·7	735·6	28·96
1 dyne cm^{-2} =	$9·869 \times 10^{-7}$	$10·20 \times 10^{-7}$	1	0·1	$14·50 \times 10^{-6}$	$2·089 \times 10^{-3}$	10^{-3}	$750·1 \times 10^{-6}$	$29·53 \times 10^{-6}$
1 Pascal =	$9·869 \times 10^{-6}$	$10·20 \times 10^{-6}$	10	1	$14·50 \times 10^{-5}$	$2·089 \times 10^{-2}$	10^{-2}	$750·1 \times 10^{-5}$	$29·53 \times 10^{-5}$
1 pound force in^{-2} =	$6·805 \times 10^{-2}$	$7·031 \times 10^{-2}$	$6·895 \times 10^4$	$6·895 \times 10^3$	1	144	68·95	51·71	2·036
1 pound force ft^{-2} =	$4·725 \times 10^{-4}$	$4·882 \times 10^{-4}$	478·8	47·88	$6·944 \times 10^{-3}$	1	$47·88 \times 10^{-2}$	0·3591	$14·14 \times 10^{-3}$
1 millibar =	$0·9869 \times 10^{-3}$	$1·020 \times 10^{-3}$	10^3	10^2	$14·50 \times 10^{-3}$	2·089	1	0·7500	$29·53 \times 10^{-3}$
1 torr =	$1·316 \times 10^{-3}$	$1·360 \times 10^{-3}$	$1·333 \times 10^3$	$1·333 \times 10^2$	$1·934 \times 10^{-2}$	2·784	1·333	1	$3·937 \times 10^{-2}$
1 barometric in. Hg =	$3·342 \times 10^{-2}$	$3·453 \times 10^{-2}$	$3·386 \times 10^4$	$3·386 \times 10^3$	$4·912 \times 10^{-1}$	70·73	33·87	25·40	1

1 torr = 1 barometric mm Hg density 13·5951 g cm^{-3} at 0°C and acceleration due to gravity 980·665 cm s^{-2}

187

11. Resistance

		abohm	*ohm*	*statohm*
1 abohm	=	1	10^{-9}	$1 \cdot 113 \times 10^{-21}$
1 ohm	=	10^9	1	$1 \cdot 113 \times 10^{-12}$
1 statohm	=	$8 \cdot 987 \times 10^{20}$	$8 \cdot 987 \times 10^{11}$	1

1 International ohm = 1·00049 absolute ohm

12. Capacitance

		abfarad	*farad*	*statfarad*
1 abfarad	=	1	10^9	$8 \cdot 987 \times 10^{20}$
1 farad	=	10^{-9}	1	$8 \cdot 987 \times 10^{11}$
1 statfarad	=	$1 \cdot 113 \times 10^{-21}$	$1 \cdot 113 \times 10^{-12}$	1

1 International farad = 0·999 51 absolute farad

13. Inductance

		abhenry	*henry*	*stathenry*
1 abhenry	=	1	10^{-9}	$1 \cdot 113 \times 10^{-21}$
1 henry	=	10^9	1	$1 \cdot 113 \times 10^{-12}$
1 stathenry	=	$8 \cdot 987 \times 10^{20}$	$8 \cdot 987 \times 10^{11}$	1

1 International henry = 1·000 49 absolute henry

14. Electric Charge

		abcoulomb	amp-hour	coulomb	faraday	statcoul
1 abcoulomb	=	1	$2 \cdot 778 \times 10^{-3}$	10	$1 \cdot 036 \times 10^{-4}$	$2 \cdot 998 \times 10^{10}$
1 ampere hour	=	360	1	3600	$3 \cdot 730 \times 10^{-2}$	$1 \cdot 079 \times 10^{13}$
1 coulomb	=	0·1	$2 \cdot 778 \times 10^{-4}$	1	$1 \cdot 036 \times 10^{-5}$	$2 \cdot 998 \times 10^{9}$
1 faraday	=	9649	26·81	$9 \cdot 649 \times 10^{4}$	1	$2 \cdot 893 \times 10^{14}$
1 statcoulomb	=	$3 \cdot 336 \times 10^{-11}$	$9 \cdot 266 \times 10^{-14}$	$3 \cdot 336 \times 10^{-10}$	$3 \cdot 456 \times 10^{-15}$	1

1 International coulomb = 0·99985 absolute coulomb

15. Electric current

		abamp	*ampere*	*statamp*
1 abampere	=	1	10	$2{\cdot}998 \times 10^{10}$
1 ampere	=	0·1	1	$2{\cdot}998 \times 10^{9}$
1 statampere	=	$3{\cdot}336 \times 10^{-11}$	$3{\cdot}336 \times 10^{-10}$	1

1 International ampere = 0·99985 absolute ampere

16. Electrical potential

		abvolt	*volt*	*statvolt*
1 abvolt	=	1	10^{-8}	$3{\cdot}336 \times 10^{-11}$
1 volt	=	10^{8}	1	$3{\cdot}336 \times 10^{-3}$
1 statvolt	=	$2{\cdot}998 \times 10^{10}$	$2{\cdot}998 \times 10^{2}$	1

1 International volt = 1·00034 absolute volt

17. Magnetic flux

		maxwell	*kiloline*	*weber*
1 maxwell (1 line)	=	1	10^{-3}	10^{-8}
1 kiloline	=	10^{3}	1	10^{-5}
1 weber	=	10^{8}	10^{5}	1

18. Magnetic flux density

		gauss	*weber* m^{-2}	*gamma*	*maxwell* cm^{-2}
1 gauss (line cm^{-2})	=	1	10^{-4}	10^{5}	1
1 weber m^{-2}	=	10^{4}	1	10^{9}	10^{4}
1 gamma	=	10^{-5}	10^{-9}	1	10^{-5}
1 maxwell cm^{-2}	=	1	10^{-4}	10^{5}	1

19. Magnetomotive force

		abamp turn	*amp turn*	*gilbert*
1 abampere turn	=	1	10	12·57
1 ampere turn	=	10^{-1}	1	1·257
1 gilbert	=	$7·958 \times 10^{-2}$	0·7958	1

20. Magnetic field strength

		amp turn cm^{-1}	*amp turn* m^{-1}	*oersted*
1 amp turn cm^{-1}	=	1	10^2	1·257
1 amp turn m^{-1}	=	10^{-2}	1	$1·257 \times 10^{-2}$
1 oersted	=	0·7958	79·58	1

21. Illumination

		lux	*phot*	*foot-candle*
1 lux (lm m^{-2})	=	1	10^{-4}	$9·29 \times 10^{-2}$
1 phot (lm cm^{-2})	=	10^4	1	929
1 foot-candle (lm ft^{-2})	=	10·76	$10·76 \times 10^{-4}$	1

1 lux = 1 metre candle

22. Luminance

		nit	stilb	$cd\,ft^{-2}$	apostilb	lambert	foot-lambert
1 nit (cd m^{-2})	=	1	10^{-4}	9.29×10^{-2}	π	$\pi \times 10^{-4}$	0·292
1 stilb (cd cm^{-2})	=	10^4	1	929	$\pi \times 10^4$	π	2920
1 cd ft^{-2}	=	10·76	1.076×10^{-3}	1	33·8	3.38×10^{-3}	π
1 apostilb (lm m^{-2})	=	$1/\pi$	$1/(\pi \times 10^4)$	2.96×10^{-2}	1	10^{-4}	9.29×10^{-2}
1 lambert (lm cm^{-2})	=	$1/(\pi \times 10^{-4})$	$1/\pi$	296	10^4	1	929
1 foot lambert or equivalent foot candle	=	3·43	3.43×10^{-4}	$1/\pi$	10·76	1.076×10^{-3}	1

Luminous intensity of candela = 98·1% that of international candle
1 lumen = flux emitted by 1 candela into unit solid angle

5. Conversion factors for SI and CGS units

Quantity	Measured in CGS	SI	A	B
Mass	gram (g)	kilogram (kg)	10^3	
Length	cm	metre (m)	10^2	
Time	second (s)	second (s)	1	
Volume	cm^3	$metre^3$	10^6	
Area	cm^2	$metre^2$	10^4	
Density*	$g\ cm^{-3}$	$kg\ m^{-3}$	10^3	
Velocity	$cm\ s^{-1}$	$m\ s^{-1}$	10^2	
Moment of Inertia	$g\ cm^2$	$kg\ m^2$	10^7	
Force	dyne	newton (N)	10^5	
Work	erg	joule (J)	10^7	
Power	$erg\ s^{-1}$	watt (W)	10^7	
Capacitance		farad (F)	10^{-9}	$8{\cdot}988 \times 10^{20}$
Charge		coulomb (C)	10^{-1}	$2{\cdot}998 \times 10^{10}$
Current	biot	ampere (A)	10^{-1}	$2{\cdot}998 \times 10^{10}$
Electric field strength		$V\ m^{-1}$ or $N\ C^{-1}$	10^6	$3{\cdot}335 \times 10^{-11}$
Electric flux		coulomb (C)	$1{\cdot}257$	$2{\cdot}998 \times 10^{10}$
Inductance	cm	henry (H)	10^9	$1{\cdot}113 \times 10^{-21}$
Intensity of magnetization		$Wb\ m^{-2}$	$7{\cdot}958 \times 10^2$	$3{\cdot}335 \times 10^{-11}$
Magnetic field strength	oersted	$A\ m^{-1}$	$1{\cdot}257 \times 10^{-2}$	$2{\cdot}998 \times 10^{10}$
Magnetic flux	maxwell	weber (Wb)	10^8	$3{\cdot}335 \times 10^{-11}$
Magnetic flux density	gauss	tesla	10^4	$3{\cdot}335 \times 10^{-11}$
Magnetic pole strength		weber (Wb)	$7{\cdot}958 \times 10^6$	$3{\cdot}335 \times 10^{-11}$
Magnetomotive force	gilbert	ampere turn (A)	$1{\cdot}257$	$2{\cdot}988 \times 10^{10}$

193

Quantity	Measured in		A	B
	CGS	SI		
Permeability		H m^{-1} or Wb A^{-1} m^{-1}	$7 \cdot 958 \times 10^5$	$1 \cdot 113 \times 10^{-21}$
Permittivity		C m^{-1} V^{-1}	$8 \cdot 854 \times 10^{-12}$	$8 \cdot 988 \times 10^{20}$
Potential difference		volt (V)	10^8	$3 \cdot 335 \times 10^{-11}$
Reluctance		A Wb^{-1}	1.257×10^{-8}	$8 \cdot 988 \times 10^{20}$
Resistance		ohm (Ω)	10^9	$1 \cdot 113 \times 10^{-21}$
Resistivity		ohm metre	10^{11}	$1 \cdot 113 \times 10^{-21}$
Surface tension	dyn cm^{-1}	N m^{-1}	10^3	
Thermal conductivity	erg s^{-1} cm^{-1} °C^{-1}	W m^{-1} °C^{-1}	10^{-5}	
Viscosity (dynamic)	poise	kg m^{-1} s^{-1}	10	
Viscosity (kinematic)	stokes	m^2 s^{-1}	10^4	
Volume susceptibility	Mx Oe^{-1} cm^{-2}	Wb A^{-1} m^{-1}	$1 \cdot 59 \times 10^{-5}$	

Column A gives the number of CGS (e.m.u.) units in one SI unit.

Column B gives the number of CGS (e.s.u.) units in one CGS (e.m.u.) unit.

The product of column A and column B gives the number of CGS (e.s.u.) units in one SI unit. The numerical values of the two columns are calculated using $2 \cdot 988 \times 10^8$ m s^{-1} as the velocity of light.

*The density of water at 4°C is 1000 kg m^{-3} in SI units.

References

Units

[1] Foster, J. C., *J. Soc. Tele. Engrs. (GB)*, **10**, 10 (1881).
[2] Weber, W. E., *Pogg. Ann. (Germany)*, **82**, 337 (1851).
[3] *Rep. Brit. Ass.* (1873) p. 222.
[4] Glazebrook, Sir Richard T., *Nature (GB)*, **128**, 17 (1931).
[5] The Federal basis for Weights and Measures. *N.B.S. Circular (USA)* No. 593 (1958).
[6] Hartree, D. R., *Proc. Cambridge Phil. Soc. (GB)*, **24**, 89 (1927).
[7] Shull, H., and Hall, G. G., *Nature (GB)*, **184**, 1559 (1959).
[8] McWeeny, R., *Nature (GB)*, **243**, 196 (1973).
[9] Kennelly, A. E., *J. Inst. Elect. Engrs. (GB)*, **34**, 172 (1905).
[10] *International Dictionary of Physics and Electronics* (London: Macmillan, 1956).
[11] Bullock, M. L., *Amer. J. Phys.*, **22**, 293 (1954).
[12] Heaviside, O., *Collected papers*, Vol. 2, p. 543 (Macmillan, 1893).
[13] Heaviside, O., *Phil. Trans. Roy. Soc. (GB)*, **183A**, 429 (1892).
[14] Kalantaroff, P., *Rev. Gen. Elect. (France)*, **16**, 235 (1929)
Kinitsky, V. A., *Amer. J. Phys.* **30**, 89 (1962).
[15] Ludovici, B. F., *Amer. J. Phys.*, **24**, 400 (1956).
[16] Maxwell, J. C., *A Treatise on Electricity and Magnetism*, Vol. II, p. 244 (Oxford, 1873).
[17] Giorgi, G. L. T., *Unita rationali di elettromagnetismo* (1904).
[18] Glazebrook, Sir Richard T., *Proc. Phys. Soc. (GB)*, **48**, 448 (1935).
[19] *Nature (GB)*, **163**, 427 (1949).
[20] Tarbouriech, M., *C. R. Acad. Sci. (France)*, **221**, 745 (1945).
[21] Goldman, D. G., and Bell, R. J., *The International System of Units* (London: HMSO, 4th edn, 1981).
[22] Henderson, J. B., *Engineering (GB)*, **116**, 409 (1923). Helmer, C. H., *Engineering (GB)*, **165**, 280 (1948).

A

[1] Cathey, W. T., *J. App. Optics (USA)*, **12**, 1097 (1973).
Hawkes, P. W., *J. App. Optics (USA)*, **12**, 2037 (1973).
[2] Lippert, B., and Miller, M. M., *J. Acoust. Soc. Amer.*, **23**, 478 (1951).

195

[3] *J. Acoust. Soc. Amer.*, **9**, 65 (1938).
[4] Kennelly, A. G., and Cook, J. H., *J. Acoust. Soc. Amer.*, **9**, 337 (1938).
[5] Webster, A. G., *Nat. Acad. Sci. Proc. (USA)*, **5**, 275 (1919).
[6] Stewart, G. W., *Phys. Rev. (USA)*, **28**, 1040 (1926).
[7] Austin-Bourke, P. M., *Irish Historical Studies*, **14**, 236 (1964/5).
[8] Fowler, R. H., and Guggenheim, E. A., *Statistical Thermodynamics*, p. 282 (Cambridge: University, 1939).
[9] *Nature (GB)*, **163**, 427 (1949).
[10] *Nature (GB)*, **78**, 678 (1908).
[11] Darwin, C., *Proc. Roy. Soc. (GB)*, **186A**, 149 (1946); *J. Inst. Elect. Engrs. (GB)*, **94**, 342 (1947).
[12] Bright, C., and Clark, L., *Rep. Brit. Ass.*, 1861, p. 37.
[13] Preece, W. H., *J. Soc. Tele. Engrs. (GB)*, **6**, 448 (1877); **7**, 84 (1878).
[14] *Rep. Brit. Ass.*, 1881, p. 425; *Nature (GB)*, **24**, 512 (1881).
[15] Joel, H. F., *J. Inst. Elect. Engrs. (GB)*, **11**, 50 (1882).
[16] Barker, G. F., *Advanced Physics*, p. 786 (London, 1892).
[17] *Trans. Inter. Union of Solar Research*, **20**, 28 (1907).
[18] Barrell, H., *Nature (GB)*. **189**, 195 (1961).
[19] *CIE Proc.*, **9**, 164 (1935).
[20] Adams, J. C., and Davis, C., *Metric Systems*, p. 147 (London, 1810).
[21] *Nature (GB)*, **111**, 101 (1923).
[22] *Physics Vade Mecum*, p. 71 (Amer. Inst. Phys., 1981).
[23] Aston, F. W., *Rep. Brit. Assoc.*, 1931, p. 333.
[24] Birge, R. T., *Rep. Progr. Phys. (GB)*, **8**, 128 (1942).
[25] Whiffen, D. H., *J. Roy. Inst. Chem. (GB)*, **84**, 133 (1960). Wichers, E., *Nature (GB)*, **194**, 621 (1962).
[26] Newlands, J. A. R., *Chem. News (GB)*, **12**, 94 (1865).
[27] Moseley, H. G. J., *Phil. Mag. (GB)*, **26**, 1024 (1913).

B

[1] Blandergroen, W., *Nature (GB)*, **167**, 1075 (1951).
[2] Bjerknes, V., *Dynamic meteorology and hydrography* (Washington: 1911).
[3] Gold, E., *Quart. J. Roy. Meteorol. Soc. (GB)*, **40**, 185 (1914).
[4] *Acoustic terminology* 1.300 (American Standards Association Z 24.1: 1951).
[5] *Conversion factors and tables*, p. 23 (British Standard 350: Part 1: 1959).
[6] *Rep. Brit. Ass.*, 1888, p. 28.
[7] Glasstone, S., *Source Book of Atomic Energy*, p. 264 (London: Macmillan, 1956).
[8] *Glossary of terms used in high vacuum technology*, Appx. B (British Standard 2951, 1958).
[9] Garbett, L. G., *Quart, J. Roy. Meteorol. Soc. (GB)*, **52**, 161 (1926).
[10] *Admiralty Weather Manual*, p. 137 (London: HMSO, 1941).
[11] *Dimensions (N.B.S. Publication, USA)*, **59**, 225 (1975).
[12] Martin, W. H., *J. Amer. Inst. Elect. Engrs.*, **43**, 797 (1924).
[13] Martin, W. H., *J. Amer. Inst. Elect. Engrs.*, **48**, 223 (1929).
[14] Polvani, G., *Nuovo Cimento Suppl.* (Italy), **8**, 180 (1951).
[15] Liehr, A. D., *Mellon Institute Theor. Chem. Preprint* No. 7 (July 1963).
[16] Nusselt, W., *VDIZ. (Germany)*, **53**, 1750 (1905).
[17] Grober, H., and Erk, S. K., *Die Grundgesetze der Warmeubertragung*, p. 185 (Berlin: Springer, 1933).

[18] American Standards Association, Sub-committee Z 10 C Mar. 1941.
McAdams, W. H., *Heat Transfer*, p. 95 (N. York: McGraw-Hill, 2nd edn, 1949).
[19] Moon, P., *J. Opt. Soc. Amer.*, **32**, 355 (1942).
[20] Hartree, D. R., *Proc. Camb. Phil. Soc. (GB)*, **24**, 89 (1928).
[21] *Rep. Brit. Ass.*, 1889, p. 42.
[22] Filon, L. N. G., *Proc. Roy. Soc.*, **83A**, 576 (1910).
[23] Davis, A. H., *Proc. Phys. Soc. (GB)*, **46**, 631 (1934).
[24] Briggs, H., *A Description of the Admirable Table of Logarithms* (London, 1616).
[25] Hanes, R. M., *J. Exptl. Physiol. (GB)*, **39**, 726 (1949).
Michaels, W. C., *J. Opt. Soc. Amer.*, **44**, 70 (1954).
[26] Joule, J. P., *Rep. Brit. Ass.*, 1843, p. 33.
[27] Powell, R. P., *Nature (GB)*, **149**, 525 (1942).

C

[1] *Conversion factors and tables*, p. 13 (British Standard 350: Part 1: 1959).
[2] *Nature (GB)*, **163**, 427 (1949).
Unit of Heat (London: Royal Society; 1950).
[3] Lodge, O. J., *Nature (GB)*, **52**, 30 (1895).
[4] Lavoisier, A. L., *Traité elementaire de la chimie* (Paris 1789).
McKie, D., and Heathcote, N. H., *Discovery of Specific and Latent Heat*, p. 38 (London: Arnold, 1935).
[5] *Rep. Brit. Ass.*, 1896, p. 153.
[6] Kohlrausch, F., *An Introduction to Physical Measurements*, p. 84 (London: 1884).
[7] Giacomo, P., *Metrologia (USA)*, **16**, 55 (1980).
[8] *J. Opt. Soc. Amer.*, **40**, 663 (1950).
[9] Violle, J., *Phil. Mag. (GB)*, **17**, 563 (1884).
[10] Waidner, C. W., and Burgess, G. K., *C. R. Acad. Sci. (France)*, **148**, 1117 (1909).
[11] *Physics in Technology (GB)*, **9**, 81 (1978).
Lighting Research and Technology (GB), **10**, 1 (1978).
[12] *Illum. Engng. (USA)*, **2**, 393 (1909).
[13] *Metropolitan Gas Act (GB)*, 23 & 24 Vict., c 125, para. 25 (1860).
[14] Kraus, E. H., and Holden, E. F., p. 99. *Gems and Gem Materials* (N. York: McGraw-Hill, 1932).
[15] Order in Council, No. 1118, 14 Oct. 1913 (London: H.M.S.O.).
[16] Walsh, J. W. T., *Photometry*, p. 5 (London: Constable, 2nd edn, 1953).
[17] *Nature (GB)*, **66**, 411 (1902).
[18] Clark, L., *Dictionary of Metric Measure* (London, 1891).
[19] Farnwell, H. W., *Amer. J. Phys.*, **13**, 349 (1939).
[20] Petroleum Production Committee, D2, p. 34 (Amer. Soc. for Testing Materials, 1956).
[21] Edgar, G., *Ind. Eng. Chem. (USA)*, **19**, 145 (1927).
[22] *Technical Standard*, No. 611, 1934 (Amer. Soc. for Testing Materials: 1934).
[23] *Petroleum Products and Lubricants*, p. 1028 (Amer. Soc. for Testing Materials: 1956).
[24] Harrison, R. D., and Thorley, N., *Nature (GB)*, **188**, 571 (1960).
[25] McGill, I. S., Menzies, D. C., and Price, M. R., *Nature (GB)*, **190**, 162 (1961).
[26] Chilton, D., *Trans. Newcomen Soc. (GB)*, **31**, 111 (1958).
[27] *The Times (GB)*, 26 Jan. 1961.

[28] Grundy, R. H., *Theory and Practice of Heat Engines*, p. 226 (London: Longmans 1942).
[29] Threshold Limit Values: Health and Safety Executive EH 15/77, (London: HMSO, 1977).
[30] Gagge, A. P., Burton, A. C., and Bazett, M. C., *Science (USA)*, **94**, 428 (1941).
[31] Pearce, E. A., and Smith, C. G., *The World Weather Guide* (London: Hutchinson, 1984).
[32] *J. Inst. Elect. Engrs. (GB)*, **94**, 342 (1947).
[33] *Nature (GB)*, **24**, 512 (1881).
[34] Shercliff, J. A., *A Textbook of Magnetohydrodynamics*, (London: Pergamon, 1965).
[35] 55 Geo, III, Ch. 94 (1815); 8 Ed. VII, Ch. 17 (1908).
[36] Thomson, J., *Rep. Brit. Ass.*, 1876, p. 33.
[37] Hofmann, A. W., *Introduction to Modern Chemistry*, p. 135 (London, 1865).
[38] Huxley, J. S., *Nature (GB)*, **180**, 454 (1957).
[39] Rutherford, E., *Nature (GB)*, **84**, 430 (1910).
[40] *Brit. J. Radiol.*, **27**, 243 (1954).
[41] *Conversion factors and tables*, p. 37 (British Standard 350: Part 1, 1959).

D

[1] Mallet, L., *Brit. J. Radiol.*, **30**, 155 (1925).
[2] Kennelly, A. E., *J. Inst. Elect. Engrs. (GB)*, **78**, 241 (1936).
[3] Wyckoff, R. D., Botset, H. G., Muskat, M., and Reed, D. W., *Rev. Sci. Instrum (GB)*, **4**, 395 (1933).
[4] Haldane, J.B. S., *Evolution (USA)*, **3**, 55 (1948).
[5] Le Fevre, R. J. W., *Dipole Moments*, p. 21 (London: Macmillan, 1938).
[6] Fairbrother, F., *Nature (GB)*, **134**, 458 (1934).
[7] *Nature (GB)*, **140**, 370 (1937).
[8] Hortin, J. W., *Proc. Inst. Radio Engrs. (USA)*, **40**, 440 (1952).
[9] Rao, V. V. L., and Lakshminaraynan, S., *J. Acoust. Soc. Amer.*, **27**, 376 (1955).
[10] Green, E. I., *Elect. Engng (USA)*, **73**, 597 (1954).
[11] Moore, J. B., *Elect. Engng (USA)*, **73**, 959 (1954).
[12] Hortin, J. W., *Elect. Engng (USA)*, **73**, 550 (1954).
[13] *Surface materials for use in radioactive areas* (British Standard 4247: Part 1: 1981).
[14] *Nature (GB)*, **66**, 411 (1902).
[15] Waters, D. W., *The Art of Navigation in England in Elizabethan and Stuart Times*, Ch. 2 (London: Hollis and Carter, 1958).
Nature (GB), **58**, 60 (1898).
[16] *Standards in Petroleum Products*, p. 157 (Philadelphia: Amer. Soc. Test. Materials, 1956).
[17] Glazebrook, R. T., *Dictionary of Applied Physics*, Vol. 3, p. 730 (London: Macmillan, 1923).
Keene, J. B., *Hydrometry* (London, 1875).
[18] Preston, W., *Correlation of hydrometer scales* (Soc. Chem. Ind.); *Density hydrometers and specific gravity hydrometers* (British Standard 718, 1960).
[19] *Standard methods for testing materials* (London: Inst. Petroleum, 1946).
[20] Merrington, A. C., *Viscometry*, p. 62 (London: Arnold, 1949).
[21] MacMichael, R. F., *Ind. Eng. Chem. (USA)*, **7**, 961 (1915).
Stanley, J., *Ind. Eng. Chem. (Anal.) (USA)*, **13**, 398 (1941).

[22] Parker, H. C., and Parker, E. W., *J. Amer. Chem. Soc.*, **46**, 313 (1924).
[23] Thomson, S. P., *Light*, p. 59 (London: Macmillan 2nd edn, 1928).
[24] Martin, L. C., *Applied Optics*, p. 34 (London: Pitman, 1930).
[25] Preece, W. H., *Elect. Rev. (GB)*, **15**, 217 (1884).
[26] Hawkes, P. W., *J. App. Optics (USA)*, **12**, 2037 (1973).
[27] Gilbert, D., *Phil. Trans. Roy. Soc. (GB)*, 25 (1827).
[28] Lardner, D., *The Steam Engine*, p. 236 (London, 1851).
[29] *Rep. Brit. Ass.*, 1873, p. 224.

E

[1] Glasser, O., *Physical Foundations of Radiology* (N. York: Harper, 1952).
[2] Richter, C. F., *Elementary Seismology*, (San Francisco: Freeman, 1958).
[3] Noyes, W. A., and Leighton, P. A., *Photochemistry of Gases*, p. 114 (N. York: Reinhold, 1941).
[4] *Proc. Phys. Soc. (GB)*, **65B**, 307 (1952).
[5] Richardson, O. W., and Compton, K. T., *Phil. Mag. (GB)*, **24**, 583 (1912).
[6] *Rev. Mod. Phys. (USA)*, **3**, 432 (1931).
[7] *Reports on the Commission on Enzymes*: Int. Union Biochem. (London: Pergamon, 1961).
[8] Gamow, G., *Nature (GB)*, **219**, 765 (1968).
[9] Barton, D. C., *Geophysical Prospecting*, p. 421 (N. York: Inst. of Mining and Metallurgical Eng., 1929).
[10] British Standard 76: 1943.
[11] *Rep. Brit. Ass.*, 1873, p. 224.
[12] Foster, G. C., *J. Soc. Tele. Engrs. and Elect. (GB)*, **10**, 19 (1881).
[13] Partington, J. D., *Text Book of Thermodynamics*, p. 521 (London: Constable, 1913).
[14] Clausius, P. J. E., *Phil. Mag. (GB)*, **35**, 406 (1868).
[15] Brockmeyer, E., *Life and Works of A. K. Erlang* (Copenhagen: 1948).

F

[1] *Encyclopaedia Britannica*, **17**, 825 (14th edn, 1946).
[2] Jenkin, F., *Rep. Brit. Ass.*, 1867, p. 483.
[3] Sabine, R., *J. Inst. Tele. Engrs. (GB)*, **1**, 246 (1872).
[4] *J. Inst. Elect. Engrs. (GB)*, **94**, 342 (1947).
[5] Waters, D. W., *The Art of Navigation in England in Elizabethan and Stuart Times*, p. 172 (London: Hollis and Carter, 1958).
[6] Hofstadter, R., *Rev. Mod. Phys. (USA)*, **28**, 214 (1956).
[7] Duggan, B. M., *Biological Effects of Radiation*, Vol. 1, p. 206 (New York: McGraw-Hill, 1936).
[8] Brown, R. H., Hazard, C., *Mon. Not. Roy. Astron. Soc.*, **111**, 365 (1951).
[9] Bickerman, J. J., *Trans. Faraday Soc. (GB)*, **34**, 634 (1938).
[10] Fry, G. A., *Illum. Engng. (USA)*, **48**, 406 (1935).
[11] Walsh, J. W. T., *Photometry*, p. 138 (London: Constable, 2nd edn, 1953).
[12] Thompson, W., and Tait, P. G., *Elements of Natural Philosophy*, p. 204 (Oxford, 1873).
[13] *Physics Today (USA)*, **10**, 34 (1957).

[14] Harper, D. R., *J. Wash. Acad. Sci. (USA)*, **18**, 469 (1928).
[15] Fourier, J. B. J., *Théorie Analytique de la Chaleur* (1822).
[16] Vernotte, P., *J. Phys. Radium (France)*, **2**, 376 (1931).
[17] Bosworth, R. C. L., *Heat Transfer Phenomenon*, p. 187 (Sydney: General Publications, 1956).
[18] Guggenheim, E. A., *Nature (GB)*, **148**, 751 (1941); *Phil. Mag. (GB)*, **33**, 479 (1942).
[19] *Applied Optics and Optical Engineering*, Ed. R. Kingslake, Vol. 5, Pt. II, p. 264 (New York: Academic Press, 1969).
[20] *Groves Dictionary of Music and Musicians*, Vol. 8, p. 763 (London: Macmillan, 1954).
[21] Brode, W. R., *Chemical Spectroscopy* (New York: John Wiley, 1939), p. 195. Bayliss, N. S., *Nature (GB)*, **167**, 367 (1951).
[22] Wrangham, D. A., *Theory and Practice of Heat Engineering*, p. 289 (Cambridge: University 2nd edn, 1948).
[23] Froude, W., *Rep. Brit. Ass.*, 1869, p. 211.
[24] *Physics Vade Mecum*, p. 183 (Amer. Inst. Phys., 1981).
[25] *Rep. Brit. Ass.*, 1876, p. 33.

G

[1] Hoggett, P., Chorley, R. S., Stoddard, D. R., *Nature (GB)*, **205**, 844 (1965).
[2] Nettleton, L. L., *Geophysical Prospecting for Oil*, p. 15 (N. York: McGraw-Hill, 1940).
[3] Glazebrook, R. T., *Dictionary of Applied Physics*, Vol 1, p. 583 (London: Macmillan, 1922).
[4] Polis, V. R., and Chai, A., *Amer. J. Phys.*, **40**, 676 (1972).
[5] Thomas, B., and Wyers, J., *Amer. J. Phys.*, **41**, 139 (1973).
[6] Weights and Measures Act. 11/12 Eliz II, Ch 31 (1963).
[7] Chree, C., *Nature (GB)*, **69**, 6 (1903).
[8] Fogel, A. I., *Qualitative Chemical Analysis*, p. 130 (London: Longmans, 1937).
[9] Mack, J. E., and Martin, M. J., *Photographic Process*, p. 207 (N. York: McGraw-Hill, 1939).
[10] Conway, E. J., *Microdiffusion, Analysis and Volumetric Error*, p. 4 (London: Crosby Lockwood, 1950).
[11] Clark, Latimer, *Dictionary of Metric Measurement* (London, 1891).
[12] *Rep. Brit. Ass.*, 1895, p. 196.
[13] *Nature (GB)*, **62**, 414 (1900).
[14] *Nature (GB)*, **126**, 252 (1930).
[15] Lennie, K. S., *Engineering (GB)*, **165**, 40 (1948).
[16] Dean, R. B., *J. Phys. Coll. Chem. (USA)*, **55**, 611 (1951).
[17] *J. Inst. Elect. Engrs (GB)*, **34**, 171 (1905).
[18] *Handbook of Chemistry and Physics*, p. 3154 (N. York: Chemical Rubber Coy. 43rd edn, 1961/62).
[19] Kaye, G. W. C., and Laby, T. H., *Tables of Physical and Chemical Constants* (London: Longmans, 1959).
[20] Price, E. W., *Amer. J. Phys.*, **25**, 120 (1957).
[21] Mills, B. D., *Amer. J. Phys.*, **27**, 62 (1959).
[22] Gray, H. J., *Dictionary of Physics* (London: Longmans, 1958).
[23] Kasner, E., and Newman, J., *Mathematics and the Imagination*, p. 23 (London: Bell, 1949).

[24] O'Connor, D. J., *Recueil Trav. chim. Pays-Bas (France)*, **75**, 938 (1956).
[25] Graetz, L., *Ann. Phys. (Germany)*, **18**, 79 (1883).
[26] American Standards Association Sub-Committee Z 10 C, Mar. 1941.
[27] Deslambre, J. S., *Base de System Metrique Decimal* (1806) Vol. 1, p. 58–61.
[28] *Brit. J. Radiol.*, **27**, 243 (1954).
[29] *Glossary of Terms used in Radiology*, p. 18 (British Standard 2597: 1955).
[30] Prandtl, L., *Essentials of Fluid Dynamics*, p. 413 (London: Blackie, 1952).
[31] Lorenz, H. H., *Zeits. f. techn. Physik (Germany)*. **15**. 362 (1934).
[32] *Dimensions, (NBS Publication, USA)*, **59**, 225 (1975).
[33] *Surface materials for use in radioactive areas* (British Standard 4247: Part 1: 1981).

H

[1] Tabor, D., *Hardness of metals* (Oxford: 1951).
[2] Mohs, F., *Grundriss der mineralogie* (1824).
[3] Ridgeway, R. R., Ballard, A. H., and Bailey, B. L., *Trans. Electrochem. Soc. (USA)*, **63**, 267 (1933).
[4] Brinell, J. A., *J. Iron Steel Inst. (GB)*, **59**, 243 (1901).
[5] *Manual of British Water Supply*, p. 533 (Cambridge: Heffer, 1950).
[6] Shercliff, J. A., *A Textbook of Magnetohydrodynamics* (London: Pergamon, 1965).
[7] Shull, H., and Hall, G. G., *Nature (GB)*, **184**, 1559 (1959).
[8] Walsh, J.W. T., *Photometry* (London: Constable, 2nd edn, 1953).
[9] *Letter Symbols for Chemical Engineering*, p. 9 (American Standards Association Y 10.12, 1959).
[10] Hedstrom, B. O. A., *Industr. Engng. Chem.*, **44**, 651 (1952).
[11] Walsh, J. W. T., *Photometry*, p. 8 (London: Constable, 2nd edn, 1953).
[12] Guggenheim, E. A., *Trans. Faraday Soc. (GB)*, **36**, 139 (1940).
[13] *J. Inst. Elect. Engrs (GB)*, **94**, 342 (1947).
[14] *J. Inst. Elect. Engrs (GB)*, **23**, 72 (1894).
[15] Barker, G. F., *Advanced Physics*, p. 816 (London, 1893).
[16] Moon, P., *J. Opt. Soc. Amer.*, **32**, 355 (1942).
[17] *Elect. Engng (USA)*, **53**, 402 (1934).
[18] Dickinson, H. W., *James Watt*, p. 145 (Cambridge: Univ. Press, 1936). Gregory, O., *Edinburgh Review (GB)*, **13**, 324 (1809).
[19] *Rep. Brit. Ass.*, 1889, p. 43.
[20] Definition of Time Act (GB), 43 & 44 Vict., c. 9 (1880).
[21] Ward, F. A. B., *Time Measurement*, p. 11 (London: HMSO, 1949).
[22] Gamow, G., *Nature (GB)*, **219**, 765 (1968).
[23] Nernst, W., *Z. Electro. Chem. (Germany)*, **5**, 253 (1900).
[24] Lewis, G. N., *J. Amer. Chem. Soc.*, **35**, 1 (1913).

I

[1] *Nature (GB)*, **183**, 80 (1959).
[2] Gamow, G., *Nature (GB)*, **219**, 765 (1968).
[3] Anderson, H. L., Fermi, E. *et al.*, *Phys. Rev. (USA)*, **72**, 21 (1947).

[4] Burn, J. H., Finney, D. J., and Goodwin, L. G., *Biological Standardization*, p. 423 (Oxford University Press, 1952).
[5] Documenta Geigy, Scientific Tables 7th edn (eds K. Diem and C. Lentner) Geigy S.A., Basle (1970).
[6] British National Formularly. British Medical Association and the Pharmaceutical Society of Great Britain. Pharmaceutical Press, London (1955).
[7] Lewis, G. N., and Randall, M., *J. Amer. Chem. Soc.*, **43**, 1140 (1921).

J

[1] Jansky, K. G., *Proc. Inst. Radio Eng. (USA)*, **20**, 1920 (1932).
[2] Ryle, M., *Rpt. Prog. Phys. (GB)*, **13**, 184 (1950).
[3] *Admiralty Handbook of Wireless Telegraphy*, p. 68; 1938, para. 168 (London: HMSO, 1920).
[4] Harris, W. S., *Phil. Mag. (GB)*, **4**, 436 (1834).
[5] *Rep. Brit. Ass.*, 1888, p. 56.
[6] *Nature (GB)*, **163**, 427 (1949); **164**, 262 (1949).
[7] *Rep. Brit. Ass.*, 1896, p. 152.

K

[1] *Glossary of terms used in radiology*, p. 19 (British Standard 2597; 1956).
[2] Vries, L. de, *German English Science Dictionary*, p. 558 (N. York: McGraw-Hill, 2nd edn, 1946).
[3] Kapp, G., *J. Soc. Tele. Engrs. and Elect. (GB)*, **15**, 518 (1886).
[4] Meggers, W. F., *J. Opt. Soc. Amer.*, **41**, 1064 (1951); **43**, 411 (1953).
[5] Bladergroen, W., *Nature (GB)*, **167**, 1075 (1951).
[6] Candler, C., *Nature (GB)*, **167**, 649 (1951).
[7] Stoney, G. J., *Rep. Brit. Ass.*, 1871, p. 42.
[8] Hartley, W. N., *Trans. Chem. Soc. (GB)*, **43**, 390 (1883).
[9] *Nature (GB)*, **220**, 651 (1968); *Metrologia (USA)* **5**, 41 (1968).
[10] *Brit. J. Radiology*, **36**, 625 (1963).
[11] Comm. Inter. du Metre: *Procès Verbaux* (1875), p. 36.
[12] Stott, V., *Nature (GB)*, **124**, 622 (1922).
[13] Weights and Measures Act. 11/12 Eliz II, Ch 31 (1963).
[14] Electric Lighting Order Confirmation Act (GB), 46 Vict., c. 216 (1883).
[15] *Rep. Brit. Ass.*, 1888, p. 27.
[16] Firestone, F. A., *J. Acoust. Soc. Amer.*, **4**, 256 (1933).
[17] Butterworth, S., *Structural Analysis*, p. 5 (London: Longmans, 1948).
[18] Waters, D. W., *Art of Navigation in England in Elizabethan and Stuart Times*, Ch. 5 (London: Hollis and Carter, 1958).
[19] Liepmann, H. W., and Roshko, A., *Elements of Gas Dynamics*, p. 381 (N. York: Wiley, 1957).
[20] Geddes, J. A., Dawson, D. H., *Industr. Engng. Chem.*, **34**, 163 (1942).

L

[1] Kirk, P. L., *Mikrochemie (Germany)*, **14**, 1 (1933).
[2] Walsh, J. W. T., *Photometry*, p. 138 (London: Constable, 2nd edn, 1953).

[3] *Nature (GB)*, **160**, 327 (1947).
[4] Linke, F., *Handbuch der Geophysik (Germany)*, **8**, 30 (1942).
[5] Glazebrook, R. T., *Dictionary of Applied Physics*, Vol. 1, p. 583 (London: Macmillan, 1922).
[6] Klinderberg, A. and Mooy, H. M., *Chem. Engr. Progr. (USA)*, **44**, 17 (1948).
[7] Lewis, G. W., *J. Roy. Aero. Soc. (GB)*, **43**, 771 (1939).
[8] Bur. Standards Sci. Paper (USA), 475, p. 174.
   Kaye, G. W. C., and Laby, T. H., *Tables of Physical and Chemical Constants*, p. 75 (London: Longmans, 12th edn, 1959).
[9] *Athenaeum (GB)*, p. 313 (March 1888).
[10] *J. Polymer Sci. (USA)*, **8**, 270 (1952).
[11] *Nature (GB)*, **65**, 538 (1902).
[12] Bigg, P. H., *Brit. J. App. Phys.*, **16**, 119 (1965).
[13] Townsend, F. H., *Nature (GB)*, **155**, 545 (1945).
[14] Rose, F. C., *Nature (GB)*, **156**, 269 (1945).
[15] Denne, E., *Nature (GB)*, **156**, 146 (1945).
[16] Baldwin, C. C., Tonks, L., *Nature (GB)*, **203**, 633 (1964).
[17] White, H. E., *Atomic Spectra*, p. 158 (N. York: McGraw-Hill, 1934).
[18] Fletcher, H., *J. Acoust. Soc. Amer.*, **9**, 276 (1937); *Rev. Mod. Phys. (USA)*, **12**, 60 (1940).
[19] *J. Acoust. Soc. Amer.*, **14**, 105 (1942).
[20] Jones, L. A., *J. Opt. Soc. Amer.*, 27, 207 (1937).
[21] Poincaré, R., *Rev. gen. Elect. (France)*, **6**, 313 (1919).
[22] Moon, P., *J. Opt. Soc. Amer.*, **32**, 355 (1942).
[23] *Trans. Illum. Engng. Soc. NY (USA)*, **20**, 629 (1925).
[24] Shercliff, J. A., *A Textbook of Magnetohydrodynamics*, (London: Pergamon, 1965).
[25] Monash, B., *Electrical World (USA)*, **54**, 1053 (1909).
[26] Troland, L. T., *Illum. Engng. (USA)*, **11**, 947 (1916).

M

[1] Mach, E., *Wien Akad. Sitzber (Aust.)*, **96**, 164 (1887).
[2] *Rev. Mod. Phys. (USA)*, **3**, 432 (1931).
[3] Feinberg, R., *Nature (GB)*, **156**, 85 (1945).
[4] Madelung, E., *Physik Zeits*, **19**, 524 (1918).
[5] Haxel, O., Jensen, J. H. D., and Suess, H. E., *Phys. Rev. (USA)*, **75**, 1766 (1949).
[6] Weiss, P., *C. R. Acad. Sci. (France)*, **152**, 187 (1911).
[7] Allen, H. S., *Proc. Phys. Soc. (GB)*, **27**, 429 (1915).
[8] Kunz, J., *Phys. Rev. (USA)*, **25**, 115 (1925).
[9] Millman, S. S., and Rabi, I., *Phys. Rev. (USA)*, **47**, 801 (1935).
[10] Millman, S. S., *Phys. Rev. (USA)*, **47**, 745 (1935).
[11] *Nature (GB)*, **62**, 414 (1900).
[12] *Nature (GB)*, **126**, 252 (1930).
[13] Richards, T. W., and Glucker, F. T., *J. Amer. Chem. Soc.*, **47**, 1890 (1925).
[14] Firestone, F. A., *J. Acoust. Soc. Amer.*, **4**, 256 (1933).
[15] Stevens, S. S., Newman, E. B., and Volkmann, J., *J. Acoust. Soc. Amer.*, **8**, 188 (1937).
[16] *Acoustical Terminology*, p. 22 (American Standards Association Z 24.1–1951).

[17] Giacomo, P., *Metrologia (USA)*, **20**, 25 (1984).
      Rowley, W. R. C., *Phys. Bull. (GB)*, **35**, 282 (1984).
[18] Babinet, J., *Ann. Chim. Phys. (France)*, **40**, 177 (1828). *Units and Standards of Measurement employed at the National Physical Laboratory, London* (HMSO, 1951).
[19] Weights and Measures Act, 11/12 Eliz. II, Ch 31 (1963).
[20] Beardsley, N. F., *Science (USA)*, **89**, 58 (1939).
[21] *Electrical Units of Measurements: Kelvin's Collected Papers*, Vol. V, p. 446 (Cambridge, 1910).
[22] Cumming, A. C. and Kay, S. A., *A textbook of quantitative chemical analysis*, p. 331 (London: Gurney and Jackson, 4th edn, 1922).
[23] *Admiralty Handbook of Wireless Telegraphy*, 1920, p. 56 (London: HMSO, 1920); 1938, para. 143 (London: HMSO, 1938).
[24] Harkins, W. D., and Roberts, L. E., *J. Amer. Chem. Soc.*, **44**, 665 (1922).
[25] Clark, L., *Dictionary of Metric Terms* (London: 1891).
[26] Parliamentary Papers (GB), 1895 (432, Appx. 10), XXXIV.
[27] *Nature (GB)*, **220**, 651 (1968).
[28] Burlington, R. S., *Amer. Math. Monthly*, **48**, 188 (1942).
[29] *Conversion factors and tables*, p. 44 (British Standard 350: Part 1: 1959).
[30] Dickson, H. W., and Rogers, H., *Trans. Newcomen Soc. (GB)*, **21**, 94 (1941).
[31] Sabine, R., *J. Inst. Elect. Engrs. (GB)*, **1**, 246 (1872).
[32] Weights and Measures Regulation No. 698 (London: HMSO, 1905)
[33] *Conversion factors and tables*, p. 13 (British Standard 350, Part 1: 1959).
[34] Waters, D. W., *Art of Navigation in England in Elizabethan and Stuart Times*, p. 423 (London: Hollis and Carter, 1958).
[35] Miller, W. H., *A Treatise on Crystallography* (Cambridge, 1839).
[36] *Glossary of terms used in radiology*, p. 59 (British Standard 2597: 1955).
[37] Priest, I. G., *J. Opt. Soc. Amer.*, **23**, 41 (1933).
[38] *The Times (GB)*, 31 Aug 1965: 15 Sept 1965.
[39] Beranek, L. L., *Acoustics*, p. 52 (N. York: McGraw-Hill, 1954).
[40] Glazebrook, R. T., *Dictionary of Applied Physics*, Vol. 3, p. 725 (London: Macmillan, 1923).
[41] Ostwald, W. F., *Principles of Inorganic Chemistry*, p. 156 (London: Macmillan, 1902).
[42] Goodwin, H. M., and Mailey, R. D., *Phys. Rev. (USA)*, **26**, 49 (1908).
[43] *International Dictionary of Physics*, p. 584 (London: Macmillan, 1956).
[44] Goldman, D. G., and Bell, R. J., *The International System of Units* (London: HMSO, 4th edn, 1981).
      Terrien, J., *Metrologia (USA)*, **8**, 35 (1972).
[45] Pohl, R., and Stockman, F., *Z. Phys. (Germany)*, **122**, 534 (1948).
[46] Pearson, W. K. J., *J. Inst. Metals (GB)*, **93**, 171 (1964).
[47] *Methods of Testing of Raw Rubber* (British Standard 1763: Part 3: 1951).
[48] Mooney, M., *Ind. Eng. Chem. (Anal. USA)*, **6**, 147 (1934).
[49] Siegbahn, K. (ed.), *Beta and Gamma Ray Spectroscopy* (Ch. 13) (Amsterdam: N. Holland, 1955).
[50] Sears, F. W., *Amer. J. Phys.*, **28**, 167 (1960).
[51] *Conversion factors and tables*, p. 45 (British Standard 350: Part 1: 1959).
      Schneider, E. E., *J. Phys. (GB)*, **10**, E, 2 (1977).
[52] *Proc. Phys. Soc. (GB)*, **65B**, 307 (1952).
[53] Stoney, G. J., *Phil. Mag. (GB)*, **36**, 138 (1868).
[54] *Groves Dictionary of Music*, **VII**, p. 438 (London: Macmillan, 1954).

N

[1] Pollard, E. C., and Davidson, W. L., *Applied Nuclear Physics*, p. 165 (N. York: Wiley, 1945).
[2] Nepero, Ioanne, *Mirifici Logarithmorum Canonis Descriptio* (Edinburgh, 1614).
[3] Martin, W. H., *Trans. Amer. Inst. Elect. Engrs.*, **48**, 223 (1929).
[4] Moon, P., *J. Opt. Soc. Amer.*, **32**, 356 (1942).
[5] Rogers, F. J., *Phys. Rev. (USA)*, **11**, 115 (1900).
[6] Robertson, D., *Electrician (GB)*, 24 Apr. 1904, p. 24.
[7] Hartshorn, L., and Vigoureux, P., *Nature (GB)*, **136**, 397 (1935).
[8] *Glossary of terms used in nuclear science and technology* (London: BS 3455: 1973).
[9] *CIE Proc.*, **11**, 145 (1948).
[10] Harris, C. M., *Handbook of Noise Control*, Ch. 36 (N. York: McGraw-Hill, 1957).
[11] Gay-Lussac, L. J., *Instruction sur l'essai des matières par la voie humide* (1833).
[12] Newth, G. S., *A Manual of Chemical Analysis*, p. 313 (London, 1899).
[13] Buckley, H., *Rep. Prog. Phys. (GB)*, **8**, 334 (1942).
[14] Kryter, K. D., *J. Acoust. Soc. Amer.*, **31**, 1424 (1959).
[15] Dirac, P. A. M., *Nature (GB)*, **139**, 323 (1937).
[16] Petley, B. W., *The Fundamental Constants and the Frontier of Measurement*, p. 41 (Bristol: Adam Hilger, 1985).

O

[1] *Nature (GB)*, **126**, 252 (1930).
[2] *Nature (GB)*, **62**, 414 (1900).
[3] *J. Inst. Elect. Engrs. (GB)*, **34**, 171 (1905).
[4] Everett, J. D., *Units and Physical Constants*, p. 139 (London, 1879).
[5] Lenz, H. F. E., *Pogg. Ann. (Germany)*, **45**, 105 (1838).
    Jenkins, F., *Proc. Roy. Soc. (GB)*, **14**, 154 (1865).
    Sims, L. G. A., *Bull. Brit. Soc. Hist. Sci.*, **2**, 1957.
[6] *Rep. Brit. Ass.*, 1873, p. 222.
[7] Bright, C., and Clark, L., *Electrician (GB)*, Nov. 1861, *Rep. Brit. Ass.*, 1861, p. 37.
[8] *Nature (GB)*, **24**, 512 (1881).
[9] *Nature (GB)*, **78**, 678 (1908).
[10] *J. Inst. Elect. Engrs. (GB)*, **94**, 342 (1947).
[11] *J. Inst. Elect. Engrs. (GB)*, **77**, 732 (1935).
[12] Preece, W. H., *Phil. Mag. (GB)*, **33**, 397 (1867).
[13] *Rep. Brit. Ass.*, 1865, p. 310.
[14] Sabine, W. C., *Amer. Architect.*, **68**, 1900 (1911).
[15] Ferguson, W. B., *Photographic Researches of Hurter and Driffield*, Ch. 5 (London: Royal Photographic Soc., 1920).

P

[1] *Sizes of papers and boards* (British Standard 4000, 1968).
[2] Spike, J. E., *Amer. J. Phys.*, **8**, 123 (1940).

[3] Crommelin, A. C. D., *Rep. Progr. Phys. (GB)*, **5**, 92 (1938).
[4] Hatch, F. H., and Rastall, R. H., Petrology and the Sedimentary Rocks (London: Thomas Murphy 4th Edn 1965).
    Pettijohn, E. J., *Sedimentary Rocks* (London: Harper and Row 2nd edn, 1957).
[5] Kaye, G. W. C., and Laby, T. H., *Tables of Physical and Chemical Constants*, p. 7 (London: Longmans, 1956).
[6] *Glossary of terms used in radiology*; p. 19 (British Standard 2597, 1955).
[7] Prandtl, L., *Essentials of fluid dynamics*, p. 401 (London: Blackie, 1952).
[8] American Standards Association Sub-Committee Z 10 C, Mar. 1941.
[9] Voce, E. H., *Trans. Newcomen Soc. (GB)*, **27**, 138 (1950).
[10] Bayliss, N., *Nature (GB)*, **167**, 367 (1951).
[11] Sorenson, S. P. L., *Biochem. Z. (Germany)*, **21**, 131, 201 (1909).
[12] Clark, W. M., *Determination of Hydrogen Ions*, p. 36 (Baltimore: Williams and Wallace, 2nd edn, 1922).
[13] Bates, R. G., *Electrometric pH Determinations*, p. 19 (N. York: John Wiley, 1954).
[14] Kingsbury, E., *Phys. Rev. (USA)*, **29**, 588 (1927).
[15] *Glossary of acoustical terms*, p. 16 (3013) (British Standard 661: 1955).
[16] Steudel, V., *Hochfrequenztech. u. Elekt. Akust. (Germany)*, **41**, 116 (1933).
[17] *Nature (GB)*, **140**, 370 (1937).
[18] Poincare, R., *Rev. gen. Elect. (France)*, **6**, 311 (1919).
[19] Moon, P., *J. Opt. Soc. Amer.*, **32**, 355 (1942).
[20] Koch, W., and Kaplan, D., *Nature (GB)*, **159**, 273 (1947).
[21] Koch, W., and Kaplan, D., *Nature (GB)*, **161**, 247 (1948).
[22] Ferguson, W. B., *Photographic Researches of Hurter and Driffield*, p. 301–341 (London: Royal Photographic Soc., 1920).
[23] Mack, J. E., and Martin, M. J., *Photographic Process*, p. 207 (N. York: McGraw-Hill, 1939).
[24] Eder, J. M., *History of Photography*, p. 452 (N. York: Columbia, 1945).
[25] *Ilford Technical Information Book*, Vol. 2 (1952)
    Speed and exposure index of photographic negative material. British Standard 1380; 1947.
[26] Bodea, E., *Giorgis rationales mks mass system mit dimensions koharnes* (Basel, 1949).
[27] Polis, V. R., and Chai, A., *Amer. J. Phys.*, **40**, 676 (1972).
[28] Thomas, E. and Wyers, J., *Amer. J. Phys.*, **41**, 139 (1973).
[29] *Determination of the viscosity of liquids in C.G.S. units*, p. 6 (British Standard 188: 1957).
[30] Deeley, R. M., and Parr, P. H., *Phil. Mag. (GB)*, **26**, 87 (1913).
[31] McIlhiney, P. C., *Ind. Eng. Chem. (GB)*, **8**, 433 (1916).
[32] Barker, G. F., *Advanced physics*, p. 645 (London, 1893).
[33] Cain, S. A., *Botanical Rev. (USA)*, **5**, 636 (1939).
[34] Weights and Measures Act., 11/12 Eliz II, Ch 31, (1963).
[35] *Nature (GB)*, **183**, 80 (1959).
[36] Bigg, P. H., *Nature (GB)*, **194**, 719 (1962).
[37] Thomson, W., and Tait, P. G., *Elements of Natural Philosophy*, Part 1, p. 220 (Oxford, 1879).
[38] Fleming, J. A., *J. Inst. Elect. Engrs. (GB)*, **21**, 606 (1892).
[39] Resnick, R., and Halliday, D., *Physics*, Appx. H (N. York: Wiley, 1960).
[40] Nusselt, W., *VDIZ (Germany)*, **54**, 1155 (1910).
[41] Prandtl, L., *Essentials of Fluid Mechanics*, p. 407 (London: Blackie, 1952).
[42] Preece, W. H., *J. Inst. Elect. Engrs. (GB)*, **20**, 609 (1891).

[43] Fanger, P. O., *Thermal Comfort* (N. York: McGraw-Hill, 1973).
[44] *Preferred values* (British Standard 2488: 1954).
[45] *Guide to the use of preferred numbers*: International Standards Organisation ISO/R17 1955 E.
[46] Van Dyck, A., *Proc. Radio Engrs. (USA)*, **24**, 159 (1936).
[47] *Preferred numbers*: International Standards Organisation ISO/R3 1953 E. *Preferred numbers* (British Standard 2045: 1953).
[48] *Nature (GB)*, **126**, 252 (1930).
[49] Witmer, E. E., *Phys. Rev. (USA)*, **71**, 126 (1947).
[50] Campbell, A., *Proc. Phys. Soc. (GB)*, **31**, 81 (1919).

Q

[1] Bayly, B. de F., *Proc. Instn. Radio Engrs. (USA)*, **19**, 873 (1931).
[2] Beatty, R. T., *Experimental Wireless (GB)*, **7**, 361 (1930).
[3] Cockcroft, J., *Proc. Inst. Elect. Engng. (GB)*, **100**, 89 (1953).
[4] *Physics Vade Mecum*, p. 8 (Amer. Inst. Phys., 1981).
[5] *Rep. Brit. Ass.*, 1889, p. 43.

R

[1] Wilson, C. W., *Radium Therapy* (London: Chapman and Hall, 1949).
[2] *Brit. J. Radiol.*, **27**, 243 (1954); *Bull. Inst. Phys. (GB)*, **22**, 302 (1971).
[3] Patterson, R., *Brit. J. Radiol.*, **29**, 353 (1956).
[4] Russ, S., *Arch., Rad. and Electrotech.*, **23**, 226 (1918).
[5] Thompson, W., and Tait, P. G., *Elements of Natural Philosophy*, p. 39 (Oxford, 1873).
[6] Darwin, C., *Nature (GB)*, **164**, 264 (1949).
[7] Cocconni, G., *Encyclopaedia of Physics*, Vol. 46, Part 1, p. 219 (Berlin: Springer, 1959).
[8] McLachlan, N. W., *Wireless Engineer (USA)*, **11**, 489 (1934).
[9] Beranek, L. L., *Acoustics*, p. 11 (N. York: McGraw-Hill, 1954).
[10] Hunter, D. M., Roach, F. E., and Chamberlaine, J. W., *J. Atmos. Terr. Phys. (GB)*, **8**, 345 (1956).
[11] Cox, R. J., Walker, J., *Proc. Inst. Elect. Engng. (GB)*, **103B**, 578 (1956).
[12] Fermi, E., *Phys. Rev. (USA)*, **72**, 21 (1947).
[13] *Lubrication (GB)*, **16**, 25 (1961).
[14] Reynolds, O., *Phil. Trans. Roy. Soc. (GB)*, **174A**, 935 (1883). *Physics Vade Mecum*, p. 183 (Amer. Inst. Phys., 1981).
[15] Bingham, F. C., and Thompson, J. R., *J. Amer. Chem. Soc.*, **50**, 2879 (1928).
[16] Curtis, L. F., and Condon, E. V., *Brit. J. Radiol.*, **19**, 368 (1946).
[17] Barker, R. E., *Nature (GB)*, **203**, 513 (1964).
[18] Villard, M. P., *Arch. Elect. Med. (France)*, **16**, 236 (1908).
[19] *Nature (GB)*, **122**, 294 (1928).
[20] Gray, L. H., *Brit. J. Radiol.*, **29**, 356 (1956).
[21] Paterson, R., *Brit. J. Radiol.*, **29**, 353 (1956).
[22] *Code of Practice for the Protection of Persons Exposed to Ionizing Radiation*, p. 77 (London: HMSO, 1957).

[23] Glasstone, S., *Source Book of Atomic Energy*, p. 505 (London: Macmillan, 1952).
[24] Rowland, H. A., *Phil. Mag. (GB)*, **23**, 257 (1887).
[25] Sawyer, R. A., *Spectra*, p. 10 (London: Chapman and Hall, 1945).
[26] Seeley, W. J., *Wireless Engng. (USA)*, **11**, 605 (1938).
[27] Curtis, L. F., Evans, R. D., Johnson, W., and Seaborg, C. T., *Rev. Sci. Instrum. (USA)*, **21**, 94 (1950).
[28] Candler, C., *Nature (GB)*, **167**, 649 (1951).

S

[1] Sabine, W. C., *Amer. Architect.*, **68**, 1900 (1911); *Acoustical terminology*, American Standards Association Z 24.1, 1951.
[2] Frederick, H. A., *J. Acoust. Soc. Amer.*, **9**, 63 (1937).
[3] Stanton, G. T., Schmidt, F. C., and Brown, W. J., *J. Acoust. Soc. Amer.* **6**, 101 (1934).
[4] Hunt, F. V., *Acoust. Soc. Amer.*, **11**, 38 (1938).
[5] Simpson, R. H., *Weatherwise*, **27**, 169–86 (1974); Simpson, R. H. and Riehl, H., *The Hurricane and its Impact* (Oxford: Blackwell, 1981).
[6] *Letter Symbols for Chemical Engineering*, p. 9, American Standards Association Y 10.12, 1955.
[7] *Proc. Inst. Mech. Engrs. (GB)*, **155**, 145 (1946); *Workshop Practice*, British Standards Institution Handbook No. 2 (1962).
[8] Whitworth, J., *Proc. Inst. Civil Engrs. (GB)*, **1**, 157 (1841).
[9] *Rep. Brit. Ass.*, 884, p. 287.
[10] *Financial Times* (London) 24 Nov 1965.
       Metric Screw Threads, ISO 3643: 1966.
[11] Ayrton, W. E., and Perrry, J., *J. Soc. Tele. & Elect. Engrs. (GB)*, **16**, 320 (1887).
[12] Terrien, J., *Metrologia (USA)*, **9**, 43 (1972).
[13] Merrington, A. C., *Viscometry*, Chap. 5 (London: Arnold, 1949).
[14] Steinberg, J. C., *Phys. Rev. (USA)*, **26**, 508 (1925).
[15] *Letter Symbols for Chemical Engineering*, p. 9, American Standards Association Y 10.12, 1955.
[16] Galbraith, W., *Extensive Air Showers*, p. 4 (London: Butterworth, 1958).
       Janossy, L., *Cosmic Rays*, p. 205 (Oxford: Clarendon, 1948).
       Rossi, B., and Greisen, K., *Rev. Mod. Phys., (USA)*, **13**, 253 (1941).
[17] Siegbahn, M., *Spectroscopy of X Rays* (Oxford, 1925).
[18] Cohen, E. R., *Atomic data and nuclear data tables (USA)*, **18**, 587 (1976).
[19] Kennelly, A. E., *J. Inst. Elect. Engrs. (GB)*, **78**, 238 (1936).
[20] *Metrologia (USA)* **16**, 55 (1980).
[21] *Glossary of terms used in radiology*, p. 19 (British Standard 2597: 1955).
[22] Kasner, E., Newman, J. *Mathematics and the imagination* (London: Bell, 1949).
       Hardy, G. H., *Pure Mathematics* (Cambridge: University 10th edn, 1963).
[23] Buckley, H., *Rep. Progr. Phys. (GB)*, **8**, 334 (1942).
       Terry, J., *Calculus for Engineers*, p. 26 (London: Arnold, 1897).
[24] Lennie, K. S., *Engineering (GB)*, **165**, 40 (1948).
[25] Worthing, A. M., *Dynamics of rotation* (London: Longmans, 4th edn, 1902).
[26] *Phillips Research Rpts.*, **6**, 251 (1951).
[27] *Relation between the sone scale of loudness and the phon scale of loudness level* (British Standard 3045: 1958).

[28] Stevens, S. S., and Davis, H., *Hearing, its Psychology and Physiology*, p. 119 (N. York: Wiley, 1938).
[29] Callou, L., *C. R. Acad. Sci. (France)*, **218**, 66 (1944).
[30] *Smithsonian Meterological Tables*, p. 205 (Washington 6th edn, 1951).
[31] Aldridge, A., J., *J. Inst. Elect. Engrs. (GB)*, **51**, 391 (1913).
[32] Hortin, W. H., *Trans. Amer. Inst. Elect. Engrs.*, **43**, 797 (1924).
      Handbook of Wireless Telegraphy, Appx. A, Vol. 1 (London: Admiralty, 1938).
[33] Judd, D. B., *J. Opt. Soc. Amer.*, **23**, 360 (1933).
[34] Troland, L. T., *J. Opt. Soc. Amer.*, **6**, 557 (1922).
[35] Bosworth, R. C. L., *Heat Transfer Phenomenon*, p. 115 (N. York: Wiley, 1952).
[36] Pogson, N., *Monthly Not. Roy. Astron. Soc. (GB)*, **17**, 13 (1956).
[37] Russell, H. N., *Astrophys. J. (USA)*, **43**, 103 (1916).
[38] Polvani, G., *Nuova Cimento Suppl. (Italy)*, **8**, 180 (1951).
[39] Halstead, G. B., *Mensuration*, p. 78 (Boston: 1881).
[40] Thomson, J., *Rep. Brit. Ass.*, 1876, p. 33.
[41] Poincaré, R., *Rev. Gen. Elect. (France)*, **6**, 311 (1919).
[42] Moon, P., *J. Opt. Soc. Amer.*, **32**, 355 (1942).
[43] Jacob, M. Z., *Tech. Phys. (Germany)*, **9**, 21 (1928).
[44] Barr, G., *World Petroleum Congress Proc.*, **2**, 508 (1933).
[45] *Determination of the viscosity of liquids in C.G.S. Units*, p. 6 (London British Standard 188: 1957).
[46] Strouhal, V., *Ann. Phys. (Germany)*, **5**, 216 (1878).
[47] Fung, Y. C., *Introduction to the Theory of Aerodynamics*, p. 163 (N. York: Wiley, 1955).
[48] Lodge, O., *Electrician (GB)*, **29**, 371 (1892).
[49] Kryter, K. D., *J. Acoust. Soc. Amer.*, **31**, 1415 (1959).
[50] Beranek, L. L., *J. Acoust. Soc. Amer.*, **28**, 833 (1956).
[51] Beranek, L. L., *Trans. bull. No. 18*, Industrial Hygiene Foundation (USA), (1950).
[52] Miller, L. N., and Beranek, L. L., *J. Acoust. Soc. Amer.*, **29**, 1169 (1957).
[53] Eisenbud, M., *Environmental Physics*, p. 108 (New York: McGraw-Hill, 1963).
[54] *Encyclopaedia Britannica*, Vol. 6, p. 63, 14th Ed. 1946.
[55] *Technical Review*, No. 3. (Copenhagen: Bruel and Kjoer, 1961); British Standard 1134: 1959.
[56] Tolansky, S., *Multiple Beam Interferometry of Surfaces and Films*, p. 46 (Oxford, 1948).
[57] Bridgman, W. B., *J. Amer. Chem. Soc.*, **64**, 2353 (1942).

T

[1] *J. Opt. Soc. Amer.*, **27**, 211 (1937).
[2] *Nature*, **163**, 427 (1949).
[3] Thompson, W., *Collected Papers*, Vol. 1, p. 100 (Cambridge, 1882).
[4] Rankine, W. J. M., *Edinburgh Phil. Trans.* (1853).
[5] *Nature (GB)*, **220**, 651 (1968).
[6] *International Temperature Scale of 1948* (London: HMSO, 1948).
      Durieux, H., *et al.*, *Metrologia (USA)*, **15**, 57 (1979).
      Hudson, R. R., *Metrologia (USA)*, **18**, 163 (1982).
      Hall, J. A., *Metrologia (USA)*, **3**, 25 (1967).
[7] *Nature (GB)*, **187**, 1077 (1960).

[8] Kurti, N., and Simon, F., *Proc. Roy. Soc. (GB)*, **152A**, 21 (1935).
[9] Casimir, H. B. G., de Hass, W. J., and de Klerk, D., *Physica (Netherlands)*, **6**, 255 (1939).
[10] Gas Regulating Act (GB), 10 and 11, Geo. V, c. 28 (1920).
[11] *Rep. Brit. Ass.*, 1888, p. 56.
[12] Joly, J., *Nature (GB)*, **52**, 4 (1895).
[13] Bosworth, R. C. L., *Phil. Mag. (GB)*, **37**, 803 (1946).
[14] White, H. E., *Bull. Amer. Ceramic Soc.*, **17**, 17 (1938).
[15] Bosworth, R. C. L., *Nature (GB)*, **158**, 309 (1946).
[16] Poincaré, R., *Rev. gen. Elect. (France)*, **6**, 311 (1919).
[17] Sears, F. W., *Amer. J. Phys.*, **28**, 167 (1960).
[18] Pierce, F. T., and Rees, W. M., *J. Textile Ind.*, **7**, 181 (1946).
[19] Danloux-Dumesnils, M., *Étude Critique du Système Métrique* (Paris: Gauthier Villars, 1962).
[20] *Limits and fits for engineering* (British Standard 1916: Part 1: 1953).
[21] Parliamentary Papers (GB), 1882, C3380 xxi.
[22] Faires, V. M., p. 379, *Applied Thermodynamics* (N. York: Macmillan, 1947).
[23] Partington, J. R., *An Advanced Treatise on Physical Chemistry*, p. 543, Vol. 1 (London: Longmans, 1949).
[24] *Glossary of Terms used in High Vacuum Technology*, p. 20 (London: British Standard 2951: 1958).
[25] Huxley, L. G. H., Crompton, R. W., and Elford, M. T., *Brit. J. App. Phys.*, **17**, 1237 (1966).
[26] Fletcher, H., and Steinberg, J. C., *Phys. Rev. (USA)*, **24**, 307 (1924).
[27] Judd, D. B., *J. Opt. Soc. Amer.*, **23**, 360 (1933).
[28] Moon, P., and Spenser, H. G., *J. Opt. Soc. Amer.*, **35**, 410 (1945).
[29] Moon, P., *J. Opt. Soc. Amer.*, **36**, 120 (1946).
[30] Troland, L. T., *Illum. Engrs. (USA)*, **11**, 947 (1916).
[31] International Union of Biochemistry Report on the Commission on Enzymes (Oxford: Pergamon, 1961).

U

[1] van Straaten, J. F., *Thermal Performance of Buildings*, p. 56 (London: Elsevier, 1967).
[2] Essen, L., *The Measurement of Frequency and Time Interval* (London, HMSO, 1973).
[3] Terrien, J., *Metrologia (USA)* **11**, 180 (1975).
[4] Giacomo, P., *Metrologia (USA)* **17**, 70 (1981).
[5] Glascombe, P., *Metrologia (USA)* **17**, 69 (1981).

V

[1] Florescu, N. A., *Nature (GB)*, **188**, 303 (1960).
Volet, C. H., *Nature (GB)*, **188**, 1017 (1960).
[2] Beranek, L. L., *Acoustics*, p. 82 (N. York: McGraw-Hill, 1954).
[3] *Industrial liquid lubricants* (British Standard 4231: 1982).
I.S.O. 3448: 1975.

[4] *Petroleum products and lubricants*, p. 269 (Amer. Soc. for Testing Materials, 1956).
[5] Bright, C., and Clark, L., *Rep. Brit. Ass.*, 1861, p. 37.
[6] Bruce-Taylor, J. P. C., *J. Inst. Tele. Engrs. (GB)*, **2**, 449 (1873).
[7] *Nature*, **24**, 512 (1881).
[8] *J.Inst. Elect. Engrs. (GB)*, **94**, 342 (1947).
[9] *Nature (GB)*, **78**, 678 (1908).
[10] Chinn, H.A., Gannett, D. K., and Morris, R. M., *Proc. Inst. Radio. Engrs. (USA)*, **28**, 1 (1940).
[11] *Acoustical terminology* 1.405, American Standards Association Z, 24.1, 1951.

W

[1] *Rep. Brit. Ass.*, 1882, p. 6.
[2] *Nature (GB)*, **78**, 678 (1908).
[3] *J. Inst. Elect. Engrs. (GB)*, **94**, 342 (1947).
[4] *Rep. Brit. Ass.*, 1872, p. 53.
[5] Stoney, G. J., and Reynolds, J. E., *Phil. Mag. (GB)*, **42**, 41 (1871).
[6] *Rep. Brit. Ass.*, 1881, p. 425.
[7] *Rep. Brit. Ass.*, 1895, p. 196.
[8] Kennelly, A. E., *J. Inst. Elect. Engrs. (GB)*, **78**, 238 (1936).
[9] Barker, G. F., *Advanced Physics*, p. 645 (London, 1893).
    Clark, L., *Dictionary of Metric Measurements* (London, 1891).
[10] Ipsen, D. C., *Units, Dimensions and Dimensionless Numbers*, p. 188 (N. York: McGraw-Hill, 1960).
[11] Weisskopf, V. F., *Phys. Rev. (USA)*, **83**, 1073 (1951).
[12] Wilkinson, D. H., A.E.R.E. (Harwell) Rep. T/R 2492 (1958).
[13] Yaglou, C. P. and Minnard, D., *Arch. Industr. Health*, **16**, 302 (1957).
[14] *Hot environment: estimation of the heat stress on a working man based on the WBGT index*. International Standard Organisation (Geneva, 1982).
[15] Houghten, F. C. and Yaglou, C. P., *American Society of Heating and Ventilation Engineers Transactions*, **19**, 531–33 (1964).
[16] McArdel, B., Dunham, W., Holling, H. E., Ladell, W. S. S., Scott, J. W., Thomson, M. L. and Weiner, J. S., *Med. Res. Council RNP*, **47**, 391 (1947).
[17] Belding, H. S. and Hatch. T. F., *Heating, Piping, Air Cond.*, **27**, 129 (1955).
[18] Givoni, B., *Man Climate and Architecture* (London: Applied Science 2nd edn, 1963).
[19] Gagge, A. P., Stolwijk, J. A. J. and Nishi, Y., *Trans. ASHRAE*, **77**, 247–62 (1971).
[20] Vogt, S. S., Candas, V., Libert, S. P. and Daull, F., *Bioengineering, Thermal Physiology and Comfort*, eds K. Cena and S. A. Clark (Elsevier, Amsterdam: 1981); *Ergonomics*, **25**, no. 4 (1982).
[21] Oleson, B. W., *Technical Review no. 2* (Denmark: Bruel and Kjaer, 1985).
[22] Kibble, B. P., *Precision Measurement and Fundamental Constants II*, eds B. N. Taylor and W. D. Phillips, Natl. Bur. Stand. (US], Spec. Publ. 617, pp. 461–64 (1984).
[23] McNish, A. G., *Physics Today (USA)*, **10**, 19 (1957).
[24] Parliamentary Papers (GB), 1895 (346) XIII, Appx. 4.
[25] Parliamentary Papers (GB), 1883 (307) XXVII, Appx. IV; 1884 (323) XXVIII.

[26] *J. Soc. Tele. Engrs (GB)*, **8**, 476 (1879).
[27] Parliamentary Papers (GB), 1914/1916 (148) XXXVII Appx. V.
[28] U.S. Bureau of Standards, Circular 31, 1893.
[29] Dickinson, H. W., and Rogers, H., *Trans. Newcomen Soc. (GB)*, **21**, 87 (1941).
[30] *Wood Screws* (British Standard 1210: 1963).
      Dickinson, H. W., *Trans. Newcomen Soc. (GB)*, **22**, 79 (1941).

Y

[1] Weights and Measures Act *(GB)*, 1878, 41 & 42 Vict., c. 49.
[2] Weights and Measures Act. 11/12 Eliz. II, Ch 31 (1963).
[3] *Nature (GB)*, **183**, 80 (1959).
[4] Thom, A., *Megalithic Sites in Britain*, Chs. 5 and 7. (Oxford: University Press, 1967).
[5] Baynes, A. W., *J. Textile Inst. (GB)*, **48**, 254 (1957); **51**, 348 (1960).
      Yarn Counts – A Universal System (Manchester: Textile Inst., 1948).
[6] *Schedules for textile testing*, 4 GP 2, Canadian Government Specifications Board, 31 Dec 1952.

Z

[1] Fairhall, D., *Russia Looks to the Sea*, p. 37 (London: Deutsch, 1971).

Index

213